Rhythm in Art, Psychology and New Materialism

This book examines the psychology involved in handling, and responding to, materials in artistic practice, such as oils, charcoal, brushes, canvas, earth and sand. Artists often work with intuitive, tactile sensations and rhythms that connect them to these materials. Rhythm connects the brain and body to the world, and the world of abstract art. The book features new readings of artworks by Matisse, Pollock, Dubuffet, Tàpies, Benglis, Len Lye, Star Gossage, Shannon Novak, Simon Ingram, Lee Mingwei, L. N. Tallur and many others. Such art challenges centuries of philosophical and aesthetic order that has elevated the substance of mind over the substance of matter. This is a multidisciplinary study of different metastable patterns and rhythms: in art, the body and the brain. This focus on the propagation of rhythm across domains represents a fresh art historical approach and provides important opportunities for art and science to cooperate.

Gregory Minissale is Associate Professor of Modern and Contemporary Art at the University of Auckland, New Zealand. He specialises in philosophical and psychological approaches to art, and is the author of *The Psychology of Contemporary Art* (Cambridge University Press, 2013).

Rhythm in Art, Psychology and New Materialism

Gregory Minissale

The University of Auckland

CAMBRIDGE
UNIVERSITY PRESS

Shaftesbury Road, Cambridge CB2 8EA, United Kingdom

One Liberty Plaza, 20th Floor, New York, NY 10006, USA

477 Williamstown Road, Port Melbourne, VIC 3207, Australia

314–321, 3rd Floor, Plot 3, Splendor Forum, Jasola District Centre, New Delhi – 110025, India

103 Penang Road, #05–06/07, Visioncrest Commercial, Singapore 238467

Cambridge University Press is part of Cambridge University Press & Assessment, a department of the University of Cambridge.

We share the University's mission to contribute to society through the pursuit of education, learning and research at the highest international levels of excellence.

www.cambridge.org
Information on this title: www.cambridge.org/9781108932912

DOI: 10.1017/9781108917216

First published 2021
First paperback edition 2022

A catalogue record for this publication is available from the British Library

ISBN   978-1-108-83141-3   Hardback
ISBN   978-1-108-93291-2   Paperback

For Danny

Contents

Colour plates are to be found between pp. 137 and 138

Figures

Acknowledgements

Many thanks are due to the University of Auckland, faculty of the arts, for providing funds for a research assistant, image copyright and reproduction fees, and travel abroad to European and American galleries, where many ideas contained in this book were hatched. I am also grateful to Professor Emeritus John Onians, University of East Anglia; Professor Gerry Cupchik, University of Toronto; Distinguished Professor Stephen Davies, University of Auckland; Professor Whitney Davis, University of California, Berkeley; and Professor David Joselit, Harvard University – all have been very generous and supportive. I would like to thank Cambridge University Press, in particular, Ilaria Tassistro, commissioning editor for psychology, and Emily Watton, senior editorial assistant for psychology, and Anitha Nadarajan, Integra-PDY India, who have been extremely helpful in editing and preparing the manuscript.

Many thanks to those artists who have generously provided reproductions of their work: Simon Ingram, Shannon Novak, Lee Mingwei and L. N. Tallur. I would like to thank the following galleries that have supported me in this project by providing images: the Len Lye Museum in New Plymouth; The Fletcher Trust Collection, Auckland; The Nelson-Atkins Museum of Art, Kansas City, Missouri; and the Tate Gallery, London.

My deepest thanks to Firuza Pastakia for some astute and thought-provoking editing. A special thanks, also, to Victoria Wynne-Jones and Sue and Rob Gardiner for reading the manuscript and offering valuable comments. Many of the arguments in this book were honed by lively discussions with my students and colleagues at the University of Auckland – thank you for your enthusiasm and insights. The work would never have seen the light of day without the patience and encouragement of my better half, Malcolm Sired.

Introduction: A Rhythmanalysis of Art

We overlook the hairline fractures in the sheen of an oil painting in order to exchange glances with a long-deceased personage. Inconvenient truths are barely registered: a ridge of impasto collects dust, an uncertain light on the surface moves as we move, briefly exposing the warp and weft of the canvas. To focus on humble materials would only remind us that the silk, hair and flesh, the mind and soul depicted, are bits of stuff slowly decaying. Several centuries of practice have turned this habit of overlooking matter into a fine art: the image prevails over its worthless material substrate.

Why do we overlook matter in this way and what happens when we don't? This is one of the key questions I pursue in this book. An answer to the first part of this question is that for many centuries it seemed a natural function of art to express eternal ideals – the divine, the soul, the mind, order and harmony, and other immutable truths. Making art has long been an exercise, implicit or explicit, in manipulating inanimate matter and controlling the chaos and contingency that undermine these ideals. Yet artists literally hold matter in their hands, replaying these kinds of conflicts at the back of their minds. In contrast, and to answer the second part of the question, in modern abstract art, wild and rude amounts of matter seem to be all there is to look at. Without order, matter comes forth to produce a direct encounter with its rhythms, textures and viscosities, offering no narrative, meaning or form in which to find comfort.

This book studies the kind of engagement that is involved in sifting through the matter in abstract art, an experience that rhythmically switches from order to disorder and back again. In many of the artworks I examine, matter appears unmodelled and in a raw state. Such work offers the tantalising notion that, however much it presents visions of rhythmic strata, dappled shadows and swarming masses, abstract art is simply pigment, oil, sand and dust – a zone, or more particularly a piece of material, left dangling, free of artistic manipulation. When the artist relaxes control of matter, and the viewer follows this relinquishment of control, there may arise a feeling of passive receptivity to the agitated patterns of matter in the work that seem to gather and disperse of their own accord. How this kind of spontaneous, involuntary and rhythmic connection arises is the stuff of this book.

Involuntary yet structured rhythms in the artist's brain and body arise in the handling of materials in this kind of artistic practice. These internal rhythms are at the same time externalised in abstract art and may be felt as rhythm by the viewer. Rhythm is an essential way in which the brain and body are connected to the world, and this is particularly so in the world of abstract art. The kind of artistic practice that enables these connections resists centuries of philosophical and aesthetic order that has elevated the substance of mind over the substance of matter. But, as I will argue here, such art effectively, and through contact with matter, in fact eliminates this 'substance dualism'.

To understand how the rhythmic entanglement of brain, body and world emerges we must rely on different levels of description: philosophical, psychological and artistic. For a good example of how artists themselves have intuitively attempted to do this, we can look to the Catalan artist Antoni Tàpies who describes his painting as

organic elements, forms that suggest natural rhythms and the spontaneous move-ment of matter; a sense of landscape, the suggestion of the primordial unity of all things; generalized matter; affirmation of and esteem for the things of the earth . . . . meditation on a cosmic theme, reflections for contemplation of the earth, of the magma, of lava, of ash . . . . [In] Buddhist meditation, they also seek the support of certain *kasinas* that sometimes consist of earth placed in a frame, in a hole in a wall, in charred matter (Tàpies in Ishaghpour 2006, 117).

Along with post-war European and American abstract art, I focus on examples of 'matter painting', which Laurence Alloway describes as 'a form halfway between painting and sculpture' (Alloway 1960, n.p.). I examine the underlying dynamics of thoughts and feelings involved in viewing this kind of art in the hope of digging deeper into what we mean by 'abstraction'. Art historian David Sylvester suggests that matter paint-ing symbolises the 'massive materiality of the physical world, the relation-ship between man and the raw materials with which he builds, the inchoate matter which is at once responsive and resistant to his will to impose a form upon it' (Sylvester 1997, 171). But rather than being merely a vehicle for the act of painting, 'the thick opaque matter of these paintings seems not only to have a life but to have lived, to have been weathered and ravaged by time' (171). And for the philosopher Martin Heidegger in *The Origin of the Work of Art*, art 'does not cause the material to disappear, but rather causes it to come forth for the very first time' (Heidegger 2002, 46).

Abstraction is often understood as a place of lucid, conceptual calm but it can also be agitated by 'a swirling viscosity, an oneiric vagueness of forms' (Gooding 2001, 89). In matter painting and other abstract works, the artwork is not entirely a finished product of intentional thought. As art historian Rosalind Krauss writes, '[T]o say that works of art are intentional objects is to say that each bit of them is separately intended' (Krauss 1981b, 6).[1] In examining the sculptures of Auguste Rodin, Krauss shows us the importance of processes of facture and how they are relaxed to allow the textures of matter to emerge, so that these sculptures are poised between organised intention and unorganised matter. It is often by relaxing rational judgement of a painting's 'meaning' that we become sensitive to the rhythms it suggests. For the philosopher Gaston Bachelard in *Earth and Reveries of Will*, artists are 'sensitized to the rhythms of matter' (Bachelard [1948] 2002, 39). In moulding matter, 'there are no more sharp edges, no more breaks. It is a continuing dream ... it is rhythmic, with a heavy rhythm that takes hold of the whole body' (107).

An obvious instance of this rhythmic connectedness is speech. For example, psychologists maintain that

our speech and our body motions exhibit wave-like characteristics that are both personal and cultural .... [T]he analysis of rhythmic entrainment and music benefits from thinking about the transformations as not happening solely inside a particular body, but happening across several, or many bodies .... The advantages of this approach are that we are discouraged from trying to look inside a particular brain/body to find the answers to the special aura that such events have, but are looking rather at the aura of the whole (Becker 2011, 65–67).[2]

Psychologists study how different brains are coupled through 'neural entrainment', as demonstrated by individuals listening to the same story. Rhythm, interval, voice modulation and story structure help to synchronise brain oscillations that not only follow speech but anticipate what might be coming next (Hasson et al. 2012, 2015). This coupling extends to the visual modality, in interpreting gestures and facial expressions, which also amplify, modulate and entrain brain oscillations so that they synchronise across subjects.

The studies I examine in later pages show that this synchronisation not only occurs between humans but can also happen between humans and films, artworks and music. This interaction and synchrony of brain, body

---

[1] Compare this with Jasper Johns, who states: 'Intention involves such a small fragment of our consciousness and of our mind and of our life. I think a painting should include more experience than a simply intended statement' (quoted in Sylvester 1997, 465).

[2] See also Thaut, who writes that 'evidence of direct frequency entrainment in rhythmic synchronisation suggests that rhythm in music can have a profound influence on the organisation of movement in time and space ... rhythmic stimulation provides a continuous time reference to the motor system' (Thaut 2005, 43).

and world have been studied in Andy Clark's 'extended mind' theory, where the author states that 'a good deal of actual thinking involves loops and circuits that run outside the head and through the local environment' (Clark 1998, 206).[3] An example of the way in which our brains rely on non-brain things in the world to function is how we use global positioning technology to navigate through the city. In his 'material engagement theory', Malafouris (2013) makes the strong point that this coupling of what is traditionally assumed to be the domain of the mind with the matter outside of it should not be understood simply as a tool for enhancing everyday cognition. This is because the qualities of the matter or material structure cause cognition itself to make new connections and this plasticity emerges from the interconnections of brain, body and world. If art helps to extend the mind in a cooperative feedback process, it is not just for performing calculations or carrying out daily tasks.

Figure 0.1 Louise Bourgeois, *End of Softness* (1967). Bronze, 18.1 × 51.1 × 38.1 cm. The Nelson-Atkins Museum of Art, Kansas City, Missouri. Acquired through the generosity of the William T. Kemper Foundation – Commerce Bank, Trustee, 2004.40. Image: John Lamberton/Nelson-Atkins Museum of Art © Louise Bourgeois/ Licensed by VAGA at Artists Rights Society (ARS), New York, NY/ Copyright Agency, 2020.

[3] Some reservations about extended mind theories and art are discussed in detail in Minissale (2013, 251–276).

Many artistic practices I examine in this book suggest that extended mind is a different kind of mind, a brain made up of matter that is in contact with matter through the hands, skin and body, coupling sensations with memories, dreams and rhythmic kinds of reverie. An example of this is given by the French novelist Pierre Loti:

Bored and annoyed by the rain, I thought to distract myself by melting a tin plate over the fire and then pouring the scalding hot liquid into a pail of water. The tin formed a sort of twisted block, a fine light silver in colour very like a lump of ore. I stared at it dreamily, for a long while (quoted in Bachelard [1948] 2002, 212).

This reverie sensitises us to the muscular yet liquid allure of Louise Bourgeois's *End of Softness* (1967). For Bachelard, 'matter is a centre of dreaming' ([1942] 1999, 52). Significantly, for him it is 'the dream state which attends the plying of matter' (3). This is an important observation because it suggests that the attraction to matter and its rhythms, which is so important in producing and viewing matter painting and abstract art, has a closer relationship to daydreaming than to detached, rational observation. This engagement with matter in its unsettled and disordered aspect affects the psychology of observation, prompting nonlinear sequences of thought and sensation. The unpredictable rhythms of matter exhaust attempts to take control of it, and instead our mind drifts into a kind of dreaming with eyes wide open, our imagination cued by the granular textures and rhythms, the twists and turns of the matter itself. This suggests that reverie can be extended and situated, that it is not all in the head.

This 'extended reverie' is more forcefully suggested by Bachelard, for whom it is crucial 'to contemplate the universe with an imagination open to the energies of matter' (278). Poetic language, he writes, 'when it is used to translate material images, becomes a veritable incantation to the forces of energy' (6). He cautions against different kinds of phenomenology, which often 'remain too "formal," too intellectual' because they objectify 'forms and not forces' (171). Studies of form, as we see in art and gestalt psychology, are 'condemned to be only psychologies of concept or structure; they are scarcely more than psychologies of the image-filled concept' (85). This 'static realism' is inferior to the 'dynamic nature of the imagination' (85) with its sources in the oneiric.

In *Rhythm, Music and the Brain*, Michael Thaut discusses how it is common for spatial images to arise in the mind while listening to music:

[S]ound durations can express extensions and distances; rhythmic and melodic contours can express images of lines and geometric figures; vertical stacks of sound can evoke pictures of multidimensional forms and layered objects. One of the most impressive and illustrative ways to study such translations can be found in the writings and works of Paul Klee (Thaut 2005, 16).

We discern rhythm primarily from movement, by recognising repeat structures (periodic structures) and through variation or differentiation. The paradox, of course, is that paintings do not move or make a sound. This is similar to the way in which music is perceived as movement even though nothing in music actually moves. It may be the case that we project motor routines onto sound experienced as pulses or constants. Intervals and changes in volume and tempo may be felt as rhythmic shifts in time and place.

Danijela Kulezic-Wilson discusses the French composer Michel Chion's theory of 'transsensorial perception', which is 'neither specifically auditory nor visual as it becomes decoded in the brain as rhythm after passing the sensory path of the eye or ear' (Kulezic-Wilson 2015, 40). The theory holds that, although the senses pick up rhythm, there is a fundamental interpretative mechanism in the brain that is able to intuit rhythm beneath the senses. What can trigger the feeling of rhythmic processes in the brain and body is an awareness of simultaneity and sequentiality, an understanding of how events or features occur or seem to affect the senses. The impression that something is moving when it is in fact static is not new.[4] There are numerous ways in which it is possible to infer rhythm in a static medium such as painting or drawing. A well-known perceptual principle, the 'law of common fate', holds that two or more lines with similar features placed next to each other will suggest that they are moving together, when compared to other details: the two backslashes in 'http://' appear to switch to the right while the colon remains stationary. This may seem self-evident but is often not made explicit enough in our 'reading' of abstract art, where such lines and patterns are far more complex. Winawer et al. (2010) show that static pictures produce motion effects in the brain, supported by sensory neurons. Summarising many of these principles, Thaut concludes:

In the broader sense, every work of art possesses rhythm. Because rhythm deals with the discernible structure of temporal organization of an artwork's 'building blocks' into an arrangement of its physical elements into form-building patterns, rhythm is one of the most important components of an artwork . . . . [R]hythm can also be transposed to visual-spatial elements, for example, by organizing patterns of deflections in lines, by patches of distinct coloring, or by arranging similarly shaped objects in spatial configurations. The rhythms of speech and the rhythms of statements and dialogues, in conjunction with movements, can express dramatic rhythms in theatrical plays. The distribution of syllables and inflection

---

[4] Johann Wolfgang von Goethe suggests that one should close one's eyes before the Laocoön and then open them very briefly to receive the overall impression: 'By this means he will see the whole marble in motion ... it is a flash of lightning fixed, a wave petrified at the moment it rushes towards the shore' (quoted in Lampert 2012, 95).

points in poetry and the distribution of elements of motion of the human body in dance are examples of rhythms in other art forms (Thaut 2005, 4).

In a painting, a line that is broken or dotted can also be felt as rhythmic, as a pulse, and lines that are repeated alongside each other can be read as vibrations. This was a common Futurist device. In order to suggest motion in a static medium such as painting or sculpture, the Futurists attempted to agitate the psychology of the observer by providing simultaneous contrasts and collocations, and multiple and clashing light sources, and by repeating the lines of objects with emphasis to suggest centrifugal and centripetal forces. The Futurists called these lines in their paintings and drawings 'force lines' (*linee-forza*).

In Umberto Boccioni's high-contrast charcoal drawing on white paper, *Muscular Dynamism* (1913), the black outlines of a nude body walking are repeated, suggesting vibrations, motion, blur and rhythmic momentum. Boccioni read the philosophy of Henri Bergson, who believed that the past, present and future dissolve into each other like musical notes. Boccioni's drawing suggests not only how the past, present and future flow, how it takes time to stretch or to walk, but also how it takes time to rhythmically drag the charcoal across the paper to get from one point of the pictorial space to another and to repeat the process. In Brian Petrie's study of Boccioni, he observes how the artist interpreted Bergson's 'duration', the sense of time flowing, as a muscular and temporal stretching forward. In Boccioni's drawing (Figure 0.2), the implication is that in order to walk with purpose towards some destination there had to have been the initiation of an impulse to extend the leg, bend the knee and push forward the torso, which we see as taking place in the present, while there is also a sense of the future, becoming manifest in muscle readiness for the body to anticipate the next step. For Bergson, the human body is 'like a moving boundary between the future and the past' (quoted in Petrie 1974, 146).

Many writers on art are sensitised to this kind of 'rhythmic seeing'. Clearly inspired by Bachelard, Mel Gooding refers to the modality of abstraction as

a kinetic representation of the world experienced as flux, as a complex of sensations in which it is impossible to hold anything still. In this thrilling place our sensorium is assailed by the teeming facts of the actual, and their poetic realisations have the flickering inconstancy of fire. Painting of this kind revels in the evanescence of the elements, in the ceaseless play of light and shadow, in the intensities of colour, in vivid creatures, in the rhythms of free dance and the dissonances of jazz. If the art of an achieved poise is a function of reverie, of daydream, then this art of the perpetual movement has its equivalence in night-time dreaming, and is characterised as a swirling viscosity, an oneiric vagueness of forms (Gooding 2001, 88–89).

Figure 0.2 Umberto Boccioni, *Muscular Dynamism* (1913). Pastel and charcoal on paper, 86.3 × 59 cm. Image: The Museum of Modern Art, New York/Scala, Florence. Out of copyright.

Sylvester also understood static images rhythmically. He describes André Masson's paintings as 'the insistent rhyming between shapes close to one another' (Sylvester 1997, 452). A depiction of hands and fingers produces a 'vigorous and systematic rhyming, which gives the picture a very rapid tempo' (452). This is consistent with the finding that, in watching a film, individuals share various synchronies in brain activity while looking at delicate hand movements (Hasson et al. 2004). The rhythmic synchronies in reading hand movements seem important. Tapping the fingers in time to a steady beat or adjusting the rate of breathing to sing along with a song are other examples, besides speech patterns, that demonstrate how the brain, body and external rhythms in the world can synchronise. Current research paradigms balk at the complexity of such a simple moment. Kelso et al. (2013) show that rhythms can become coordinated or synchronised across multiple levels of organisation from the microbiological to the phenomenal; for example, finger tapping is produced by the brain's own rhythms in synchrony with external beats.

Sylvester may have been inspired to make his observation of the rhythmic qualities of fingers and hands by noting how Jasper Johns described his own painting, *White Flag* (1955) (Figure 0.3), as a change in rhythm: 'The change has two speeds. In the stars it's *allegro vivace*, agitated movement, flickering and exploding. In the stripes it's *andante*' (Sylvester 1997, 464). How can such a remark be understood as more than simply a metaphor, as a phenomenon that is experienced as rhythm, even though the fixed image does not move? Cotter et al. (2017) show how curved shapes provide certain rhythmic pleasures. Kim and Blake (2007) find that motion-sensitive areas of the brain are activated in abstract paintings that suggest motion. This helps to explain how some abstract paintings are often felt to be rhythmic. Bar and Neta (2006) suggest that angular and jagged edges are often associated with threat and agitated rhythms, while rounded edges help to produce comforting feelings and soothing rhythms. These responses have to do with haptic sensibilities, intertwined with rhythmic and emotional registers. There

Figure 0.3 Jasper Johns, *White Flag* (1958). Encaustic and mixed media, 198.9 × 306.7 cm. Digital Image © Private Collection/ Christies Images/Bridgeman Images © Jasper Johns/ARS. Copyright Agency, 2020.

are pleasurable sensations to be explored with continuous smooth shapes and intervals, texture and surface. Similarly, the eye tends to stop and sample the edges in irregular 'sharp objects' with jagged edges, creating jerky rhythms. And a group of small stars could be subliminally perceived as little pinpricks or shimmering points of light. Johns's own understanding of the 'two speeds' also switches from the stars to the stripes and back again, which suggests a complex dynamic complicated by the tendency for eyes to search the visual field in iterative, rhythmic movements both for large-scale scenes and in discerning rhythm in the handling of paint. In *White Flag*, the encaustic technique of mixing oil and wax reveals an intricately patterned surface on close inspection.

We engage with the 'higher' conceptual level when we make an analogy with musical terms, but this analogy is felt and experienced and is not just a turn of phrase. The terms describe motor sensations and muscle memory, which are experiential. In addition, the artistic process itself is a way of making explicit the properties of different kinds of matter. The technique of encaustic is painstaking: Johns prepared the ground of three panels with beeswax, building up multiple layers with a collage of newsprint and shreds of fabric cut out for each star and stuck onto the surface. He then dipped the panels into molten beeswax, applying pigments with more quantities of beeswax. The material process creates an overall effect where the flat image is transformed into a textural, sculptural surface. A simple design gives way to a densely patterned complexity, suggesting arrested contingency, change or decay. The fluidity of the medium dries and hardens, capturing the brushstrokes and waxy drips in a frosty white sculptural field. Scraps of newsprint suggest everyday moments and political events caught or suspended in the body of the flag. As one invests time to discover the rich detail, the artwork is transformed from an image (the American flag) into a sculptural and textural phenomenon. This experience also involves appreciating the labour-intensive aspect of the work and the temporally extended nature of perception (rather than instant image recognition). The fine-grained surface of the work is chaotic, exceptionally full of information, with uncountable marks, grooves, blotches and drips. In viewing the surface, which is the result of meticulous activity, one becomes sensitised to this 'material' and technique, and the mental image of the flag is lost. The work flickers between image and matter: it is both a timeless flag and an expressionist painting. The observer uses prior knowledge about the flag to imagine its colours, as if present perception is haunted by a memory of how the flag once was, perhaps even how it will continue to be, as a kind of museum artefact. We are mind wandering: looking back and looking forward, looking at the work while examining our own feelings, becoming lost in pipedreams. Johns is said to have had a dream about creating this piece. As with so many of his works, *White Flag* reflects back various conditions of viewing at the same time as it suggests phases of the existence of matter: the flag seems veiled with mould or glowing in the moonlight, embalmed and powdery or translucent and creamy; whichever 'phase' we settle on momentarily, the longer we look, the more associations spring to mind. The white flag is the conventional sign for a truce, but it also signifies whitewashing and erasure.

One thing that is important in enriching the experience of this work is an appreciation of the medium: molten wax – the fluidity of events in the wider world and mind, a complexity of durations – suddenly 'turning to stone'. The artist Robert Smithson's ideas provide a new perspective on this kind of abstraction:

[T]he mind and things of certain artists are not 'unities,' but things in a state of arrested disruption . . . no materials are solid, they all contain caverns and fissures. Solids are particles built up around flux, they are objective illusions supporting grit, a collection of surfaces ready to be cracked. All chaos is put into the dark inside of the art (Smithson 1996, 106–107).

It is motor imagery, a sense of rhythm, detecting texture and gesture, that allows painting to be felt as a series of complex fine- and coarse-grained vibrations across brain, body and world. Complementing this observation are the empirical studies I discuss in Part II, which reveal how the brain functions not as an ordered and linear set of processes but as a system of numerous unstable networks, where neurons are jittering on the edge of chaos, continuously ready for and acting in concert with spontaneous events in the external environment. These studies emphasise that the brain is a dynamic system of neural oscillations that travel across and are supported by particular brain areas, making functions (thoughts) richly intertwined over time. This approach is particularly appropriate as a description of what happens when one is viewing artworks over temporally extended periods.

Why Matter? Why Now?

Recent texts in philosophy and art history attest to a burgeoning interest in matter and materialism. Yet there seems to be a dearth of serious, detailed discussion of the brain and the biology of the body as part of this 'new materialism'. Similarly, there have been many recent texts in psychology and neuroscience on art, but none of these addresses the topic of matter and materiality that is of central importance to art.[5] In other words, what has not been joined up is the matter of the brain and the matter in art. Mind wandering is a form of abstraction involved in both the production and reception of abstract art. Mind wandering – involuntary non-logical thought – creates rhythmic connections between abstract art and abstract thought. This path was opened up by Bachelard in his works on the 'imagination of matter'. One of Bachelard's major objectives was 'to match the psychological study of reverie with the objective study of the images that entrance us' (Bachelard [1938] 1987, 107). He was interested in mapping the non-dualist substance that joins mind and matter through feeling, non- or semi-conscious thought, the oneiric and reverie – aspects of psychology that are studied today as 'mind wandering' and which are attuned to rhythms in art – a spell which is broken by the interference of logical and analytical thought.

Over the last decade there has been a renewed interest in understanding materialism in art. This is reflected in the many different readings in *Materiality*, edited by Petra Lange-Berndt (2015). Similarly, *A Questionnaire on Materialisms*, a recent issue of *October* (2016), contains forty-one responses from artists, curators and theorists writing about what 'new materialism' means to them. However, in all but one of these texts, consideration of the physical processes of the brain and the matter of the brain itself is left out of the equation.[6] This is the case with new materialism more generally, as we see in Coole and Frost (eds.), *New Materialisms* (2010); Dolphijn and Van Der Tuin, *New Materialism* (2012); Barrett and Bolt (eds.), *Carnal Knowledge* (2012); and Cox et al. (eds.), *Realism Materialism Art* (2018). It appears that what is needed is a new kind of 'neuromaterialism'. In this book I attempt to suggest what this 'neuromaterialism' might look like.

The anthropologist Tim Ingold is concerned that matter has been swept up into abstraction, whether Marxian, semiotic, philosophical or cognitivist:

---

[5] These are discussed mainly in Parts II and III. A note here on disambiguating terms. Materiality refers to the study of materials in various disciplines, from archaeology to philosophy and the sciences. Materialism, and historical materialism in particular, refers to the theory of how materials and the way they are exploited affect social, political and economic structures, even consciousness. Matter is used to examine key phenomena that may be compared in the domains of brain, body and art, even if these emerge in each domain in different ways. For example, the matter of rhythm, metastability (between order and chaos) and complexity are examined in each domain in order to reveal interactions. I do not suggest that matter amounts to the same thing underlying all domains.

[6] Only Caroline A. Jones seems to understand the bigger picture that I aim for here: 'Cranial grey matter, white matter, glia, and those billions of other neurons distributed throughout the body (more numerous in the gut than in the brain) fully participate in this thing we call *thinking*' (Jones 2016, 61).

[M]aterials appear to vanish, swallowed up by the very objects to which they have given birth. That is why we commonly describe materials as 'raw' but never 'cooked' – for by the time they have congealed into objects they have already disappeared. Thenceforth it is the objects themselves that capture our attention, no longer the materials of which they are made (Ingold 2007, 9).

In anthropology and archaeology, materiality is meant to explain how social systems of interaction affect and are affected by materials. But, as Ingold notes, such engagements are not

with the tangible stuff of craftsmen and manufacturers but with the abstract ruminations of philosophers and theorists .... . What academic perversion leads us to speak not of materials and their properties but of the materiality of objects? It seemed to me that the concept of materiality, whatever it might mean, has become a real obstacle to sensible enquiry into materials, their transformations and affordances (Ingold 2007, 2).

Somewhat opposed to this view is the archaeologist Christopher Tilley, who writes:

The concept of materiality is required because it tries to consider and embrace subject–object relations going beyond the brute materiality of stones. I am going beyond an empirical consideration of the stone to consider its meaning and significance. In doing so, I move from a 'brute' consideration of material to its social significance. This to me is what is meant by the concept of materiality (Tilley 2007, 17).

This view is supported by Daniel Miller (2007), an anthropologist specialising in material culture, who suggests Ingold may be prone to primitivism, positing an ideal kind of materiality untainted by modern industrial processes. It is interesting that these lively and informative discussions of materiality tend to revisit the age-old dualism between 'brute' materialism and linguistic and sociopolitical systems.

Other tendencies within new materialism attempt to overcome this dualism. An influential text for artists and art theorists is Jane Bennett's *Vibrant Matter*, where the author suggests that 'every thing is entelechial, life-ly, vitalistic' (Bennett 2010, 89). She explains that even the trash is alive:

[W]hat if the swarming activity inside my head was itself an instance of the vital materiality that also constituted the trash? ... [I]t is easy to acknowledge that humans are composed of various material parts (the minerality of our bones, or the metal of our blood, or the electricity of our neurons). But it is more challenging to conceive of these materials as lively and self-organizing, rather than as passive or mechanical means under the direction of something nonmaterial, that is, an active soul or mind (Bennett 2010, 10).

Bennett's preferred principle is a vital materialism that flows through everything. But this is another case of the missing brain and its dynamics:

how do the actual particles across the different domains swarm? How do the activity in the head and in the trash intermingle beyond a metaphor? More details are required to describe how these domains come together materially.[7]

There are various studies that attempt to integrate scientific theories and advances in our understanding of matter with approaches in the humanities, such as Katherine Hayles's *Chaos Bound* (1990) and her more recent *Unthought* (2017), which attempts to outline how non-conscious somatic and sensorimotor engagements with objects and materials constitute knowing beyond so-called detached reasoning. Karen Barad in *Meeting the Universe Halfway* maintains that matter is not a fixed entity but is 'agentive' and that it is 'produced and product-ive, generated and generative' (Barad 2007, 137), a continuous reality that smooths out traditional dualisms. Donna Haraway (1997) critiques another important dualism: male/female, the gender essentialism inher-ent in traditions of knowledge and in modern science and technology, where form and the active intellect are equated with the male principle while matter and passivity are associated with the female. Meanwhile, Johanna Drucker reminds us that

Aristotle divided the world of substance into matter and form, the stuff things are made of and the patterns or shapes they assume or are given. The philosopher aligned each with gendered attributes, matter, and material, with *mater*, the mother, and form with the paternal or masculine. These categories passed into European thought by way of those great translators and exhaustive compilers (Drucker 2009, 10).

This can be seen, for example, in Aristotle's text *On the Generation of Animals*, where he suggests that the father is form (sperm) and the mother (maternal blood) is formed by the sperm and that matter desires form 'like the female desires the male' (quoted in Didi-Huberman 2006, 210).

Haraway's text is situated in the poststructuralist tradition that reveals the fallacy of the dualism between the material and the semiotic, showing how both are implicated in structuring the world and our categories of knowledge. This is entirely in keeping with the philosophers Gilles

---

[7] Ian Buchanan suggests that the explanatory gap in Bennett's understanding of what an assemblage is (that is, how things are connected) is an ethical problem because it describes assemblages of materials or things but

does not say anything more here about the nature of the relationships between these items and the street, or the weather, or indeed herself, save the way they caught her eye. ... Bennett attributes her perception of these objects to a happy combination of her own perceptual openness and what she calls the 'thing power' of the objects themselves (Buchanan 2016).

Deleuze and Félix Guattari's attempt to show how speech, text, intellect and sensation are different 'movements' of matter. As John Marks writes:

Deleuze and Guattari's commitment to materialism means that they eschew the rigidities of linguistic structuralism in favour of analyses of the way in which systems – biological, physical, social – demonstrate self-organizing tendencies that cannot be traced back to the agency of specific individuals or components within the system (Marks 2006, 13).

In *Political Affect* (2009), John Protevi studies the 'political physiology' of brains, bodies and social relationships. Here, political messages are not only linguistic and semiotic but also affective, modelling physiognomic responses. This approach complements Manuel DeLanda's *Intensive Science and Virtual Philosophy* (2002), which proposes a holistic, entangled view of material objects, milieus, and technological and linguistic systems and brings together different kinds of scientific processes studied by complexity theory and in Deleuze and Guattari's philosophy.[8]

With all these different understandings of matter, we might be in danger of losing sight of matter itself. As Ernan McMullin sagely warned us many years ago:

It is not as though there is an absolute constituent or aspect of the world called 'matter', which different philosophers have described in ways more or less apposite. This would do if we were talking of hydrogen or nebulae. But for a 'second-level' notion, a somewhat different account is required. The 'matter' the philosopher singles out is constituted (not ontologically, of course, but epistemologically) by the conceptual system that defines it. Does this mean that there are as many 'matters' as there are different systems? This would be the opposite error. It is possible for different philosophies to focus upon the 'same' principle. But this identity is something to be carefully established, not casually assumed on the basis of an accidental similarity of term or of problem (McMullin 1963, 1).

The common principle in all these treatments of matter, and one which I will pursue in my investigation into the matter in art, is that matter should be understood as a dynamic complexity shared by the brain, body and art. This requires a rejection of rigid dualisms such as mind/matter, form/matter and active/passive, in favour of 'matter-movement', a term used by Deleuze and Guattari, along with 'matter-energy' and 'matter-flow'. Deleuze understood 'rhythm as matter' (Deleuze 2003, 107), which suggests that matter is not simply to be understood as something

---

[8] DeLanda (2002) attempts to show the links between Deleuze and Guattari's philosophy and dynamic systems theory. Protevi (2001) looks at the accompanying notions of hylomorphism and self-organisation in the history of philosophy. Bonta and Protevi (2004) treat Deleuze and dynamic systems theory with regard to its potential for geographical concepts. Bell (2006) follows up on this work, suggesting how dynamics systems theory and chaos in the brain might work for philosophical method.

that fills in an outline or form that is absolutely fixed and discrete; rhythm passes through things or things elaborate rhythm with their own activity to create an ecology of things vibrating together. The aim in this book is to provide some grasp of this wider ecology, which includes the rhythms of the brain, by looking to art and its varied materials, following the 'flow of matter', the Deleuzoguattarian phrase that has moved many of these authors. It is important to be more specific in our objectives with the help of neuroscientific, phenomenological and aesthetic approaches: to follow the flow of matter as it circulates through brain, body and world.

What this short literature review has revealed is one significant omission: an avoidance of engaging rigorously with the 'rhythm-matter' of the brain. In these works, what seems most difficult to acknowledge is that the brain is not just a meaning-making machine that pulls rank over matter: there are many intervals during which the brain rests in itself as disorganised matter. Bergson asserts that, in these intervals, 'pure perception, which is the lowest degree of mind – mind without memory – is really part of matter, as we understand matter' (Bergson [1896] 2004, 297). It is art that helps us to intuit this.

Monica Wagner's essay 'Material' (2015) offers a useful summary of the etymological roots of matter and materiality. However, artworks are never reducible to terms traced etymologically – they are not simply illustrations of statements or concepts. Such a notion ignores the many unspoken uses of matter in art that are messy and awkward, or even dirty, and which do not conform to such enunciations. What would an alternative art history consist of? It might trace the 'illogical' human-beast metamorphoses of antiquity; the countless apotropaic symbols, zoomorphic talismans and amulets that still affect superstitious minds the world over; the medieval marginalia that sprout marvellous and hideous menageries of mutant buffoons; the viral proliferation of abject objects, orifices and excreta catastrophically thrown together in the abyssal pictures of Hieronymus Bosch. Not far from this broader understanding of matter are Matthias Grünewald's suppurating wounds of Christ and the dark matter of Francisco Goya's nightmares. These examples point to a very different psychology of matter compared to what the philosophers discussed. This matter is convulsive, affective, infectious and on the edge of chaos. In these examples, we see matter imagined in a way that is used to destabilise rigid boundaries. Here, matter possesses an oneirically charged morphing power over hand–eye coordination: sometimes a surfeit of technical finesse bursts into spontaneous improvisation to produce vibrating lines, restless shapes and forms, mass and viscous elasticity. This kind of feverous creativity is inspired by images of lava, stormy seas and avalanches at the back of the mind. It is this quality of matter, formless because unsettled, evading the grasp but registering rhythmically, which unleashes an unpremeditated artistic dexterity.

Bachelard writes: 'If our dreams are monsters, it is because they translate energies' (Bachelard [1948] 2002, 79). These energies are undoubtedly sensual and rhythmic, stretched by the torsion of affects and drives, twisting away from rational analysis. This psychology of matter provides for much of the brain's metastability, allowing it to inform artistic practices. This happens in complex improvisations inspired by mental images of matter's spontaneity, or artists have more passively allowed matter to 'remain' in its raw state within their completed works. The artist Robert Morris describes this process, where sculptors left arbitrary traces of the facture of their bronzes to suggest contours and to reveal the qualities of the medium:

The visibility of process in art occurred with the saving of sketches and unfinished work in the High Renaissance. In the nineteenth century both Rodin and Rosso left traces of touch in finished work. Like the Abstract Expressionists after them, they registered the plasticity of material in autobiographical terms. It remained for Pollock and Louis to go beyond the personalism of the hand to the more direct revelation of matter itself . . . . [T]he employment of gravity and a kind of 'controlled chance' has been shared by many since Donatello in the materials/process interaction. However it is employed, the automation serves to remove taste and the personal touch by co-opting forces, images, and processes to replace a step formerly taken in a directing or deciding by the artist (Morris 1993, 44).

This notion of co-opting forces allows matter to have more of a role in deciding the outcome of the work and reveals the nature of matter rather than simply the intention of the artist. The artist allows areas of the work to remain unfinished within or alongside the more finished areas, reflecting the struggle between the two psychologies: deliberate and less so, composition and non-composition. The work of art preserves the external signs of this tension, which can be perceived by the viewer.

Even with the meticulous illusionism of traditional painting, which is far from the spontaneity of abstract art, it is possible to experience the effects of visual composition and non-composition. For example, in Giovanni Bellini's *St. Francis in the Desert* (c. 1480), the schist and strata of the rocks, which take up nearly all the pictorial space, can trigger involuntary experiences of staggered rhythms and mind wandering. Similarly, the depiction of clouds and earth in art can provide relief from following or constructing a narrative about the whole painting. Such intervals help to produce moments of daydreaming associated with mind wandering. In *Earth and Reveries of Will*, Bachelard writes: 'Quite often dreamers see heaps of rocks in cloud-filled skies' (Bachelard [1948] 2002, 142).

This is something we can also experience with Albrecht Dürer's *Tall Grass, Great Piece of Turf* (1503), a remarkable work of observation executed with subtle gradations of colour and opacities of pigment within a densely structured network of lines, which strike us as precious and skilful patterns of growth in nature. And yet the artist seems to suggest something which is far from the world of skill: a humble sod of earth, a patch of dirt and soil, blades of grass criss-crossing each other with careless abandon. The viewer alternates between the two psychologies of observation: from the daylight consciousness of the artist's technique, following the long stems that end in tight buds at the top of the painting, we lower our eyes to the earth where we imagine its depth and experience sensations of lush grass and dark soil. The artist might have experienced circular reveries and abstractions, wandering away from and back to the precise dexterity required for the execution of this work. In all these examples, whether in a deliberate attempt to imitate matter in its raw state or by glimpsing matter in its raw state (a knotted skein of paint catching the light), the mind is released to drift in wordless reverie.

The 'base' carved by Bernini to resemble a cloud in *The Ecstasy of St. Teresa* (1647–1662) retains some of the roughness of the marble before it is refined. To the observer who pays close attention to this work, it is not only the ripples in the saint's cloak and the jagged edges of the clouds that create breaks and rhythmic striations, but also the jagged matter of the marble itself, which the artist has preserved. Bernini refrains from over-working the marble, allowing its irregular granularity to remain discernible, while the fluttering robe reveals a higher level of completion leading up to the ideal 'luminosity' of the face. There is a hierarchy of refinement here, looser and coarser at the base, tighter and more polished above. It is remarkable that the form of the figure conveys thought, as does the relative formlessness of the coarser marble. St. Teresa's ecstasy seems to vibrate through the posture of the body and the fluttering of her robes, increasing in arbitrariness with the cloud petrified in stone.

There are many sculptures by Rodin where such a deliberately 'unfinished' state of coarse 'shapelessness' is evident rather than subliminal. The presence of 'unfinished' matter in artworks is even more prominent in the matter paintings of Jean Dubuffet, Jean Fautrier and Antoni Tàpies, who magnify this presence to attract reverie. Daydreaming tendencies can be prompted by the physical qualities of the matter itself. Bachelard describes the general effects that a sculpture in wood can produce:

A dialectic of torsion and free flow can be seen in the play of dark wood and light . . . . it offers a *portrait of energy* . . . . Hard matter is contemplated *dynamically* as a 'kernel of resistant matter' in 'a base of soft substance' .... Such a *material tableau* speaks to the dynamic being within the worker. Through our material and dynamic imagination, we experience the knot's external form awaken in us an internal force .... Perception is best described phenomenologically as movement *towards* .... Thus the material imagination *engages* us dynamically (Bachelard [1948] 2002, 40).[9]

This understanding of matter, which artists have been familiar with for centuries, can also be found in the domain of philosophy, particularly in the work of Gilbert Simondon who challenged the traditional presuppositions of 'hylomorphism'. Steven Shaviro explains:

Hylomorphism presumes that materiality, or the 'sensible' (that which can be apprehended by the senses alone), is passive, inert, and intrinsically shapeless, and that it can only be organized by an intelligible form that is imposed upon it from outside, or from above. Simondon argues that hylomorphism, with its rigid dualism, ignores all the intermediaries that are at work in any actual process of formation or construction. In fact, matter is never entirely passive and inert, for it always contains incipient structures. Matter already contains distributions of energy, and potentials for being shaped in particular directions or ways (Shaviro 2012, 52).[10]

Particularly interesting here is the emphasis on the dynamism of the imagination (nonlinear, oneiric, daydreaming) working with the dynamism of the sculptor's action and the lines and knots (which also appear rhythmic, dynamic and nonlinear) to give form to the imagination (rather than the imagination giving form to the inert materials). It is this confluence of psychology, the attributes of matter itself and descriptions of artistic creativity as relinquishing conscious control that find affirmation in recent research into mind wandering.

Another way to think about this kind of 'letting be', the 'release' of form (along with its cognates, formal and formalism) into sensation and reverie, is to turn to Heidegger's term *Gelassenheit*, or 'deferred-willing', meaning to allow things to appear in their own way without forcing them. Heidegger was also interested in thinking outside the distinction between activity and passivity, cultivating the 'willing of non-willing'

---

[9] Similarly, Deleuze and Guattari write:

 [I]t is a question of surrendering to the wood, then following where it leads by connecting operations to a materiality, instead of imposing a form upon a matter: what one addresses is less a matter submitted to laws than a materiality possessing a nomos [a principle]. One addresses less a form capable of imposing properties upon a matter than material traits of expression constituting affects (Deleuze and Guattari 1988, 408).

[10] Shaviro also points out that it was Bergson who allows for sentience without reflexivity, 'a kind of experience that remains "in itself" ... though immanently coincides with matter' (Shaviro 2018, 197).

(Heidegger 2010, 33).[11] This could be described as going through the exertions of mindfulness in order for a 'will-less' mind wandering to occur. We might imagine this as counting one's breath in meditation, counting sheep when trying to sleep or counting blades of grass. Rather than distancing us from the artwork, it helps to ease the passage from voluntary to involuntary thought.

In *The Origin of the Work of Art*, Heidegger discusses how art reveals the struggle between the earth and the world. As John McCumber writes, for Heidegger the work of art discloses matter in particular ways:

> In a work of art, the unique qualities of its materiality – its media – are not overcome but highlighted. It is part of the business of a statue to bring forth the striations in its stone or the grains in its wood, of a painting to bring forth the colors in its pigments, of a poem to show the semantic, as well as the auditory, beauty of its words. The work of art treats matter not as something passive and receptive but as something dynamic and configurating .... The form's action on the matter in a work of art is, therefore, not geared to materializing the form itself; it is supposed to bring the matter forth in its own uniqueness (McCumber 2013, 169).

This principle, that form can have a role in revealing matter's own uniqueness, can be seen consistently in matter painting and abstract art. For Heidegger, the world, while resting on the earth, strives to manipulate it. The world is an opening mechanism: it cannot allow or endure anything that is closed, and this is related to the kind of rational thought that we often impose on artworks. On the other hand, 'the earth, as both sheltering and concealing, tends always to draw the world into itself, and keep it hidden there' (Heidegger 2002, 26). It is only by 'letting it be' – *Gelassenheit* – that we can come closer to intuiting its contours. An interesting way to think about art is Heidegger's notion that the world is impatient and tries forcefully to open up the earth, while the earth is patient, where patience is important for 'letting it be'. I would add to 'letting it be', 'letting it happen'. But observing art also lets it happen *to us*. In art history, something of this kind has been given explicit form in Krauss's *Passages in Modern Sculpture*, where she suggests that an attitude of 'humility' (Krauss 1981a, 283) should allow a passage to open between the viewer and the artwork. A somewhat more existential nuance is provided by Timothy Leary in *High Priest* (1968):

---

[11] There is an obvious relation to quietism and Heidegger's awareness of Meister Eckhart's use of the term *Gelassenheit* in the context of renunciation of self-will, leaving matters up to the will of God, although 'Heidegger explicitly seeks to distance his thought from any deferential obedience to a divine will' (Davis 2010, xi).

Existential means you study natural events as they unfold without prejudging them with your own concepts. You surrender your mind to the events. Transactional means you see the research situation as a social network, of which the experimenter is a part. The psychologist doesn't stand outside the event but recognizes his part in it (quoted in Belgrad 1998, 150).

It seems that deferred-willing facilitates a feeling of being one with an artwork. This involves *feeling* relaxed with messy things, being absorbed in what we might call arbitrary movements of matter. This relaxation emerges from observing matter or letting it take its course, rather than micro-managing its behaviour, which would arrest its concretion over time. Patience allows matter the time to develop. In his own way, Dubuffet explains this psychology when he describes his working with matter as a kind of partnership that allows the 'accidents' and qualities of matter to remain in the finished work. This should not be understood as the artist actively expressing a mood so much as passively allowing the patterns of matter-movement to remain as they are, as a record of the artist observing them because of the way they affected his or her mood. Importantly, art in this sense is not the construction of an expression but rather the avoidance of construction, a deconstruction of the expression of anything prior. To see art in this way requires a change in the traditional understanding of expression as a movement crossing the internal domain of the artist to become external signs lodged in the artwork. Rather, expression becomes the preservation of the arbitrary movement and consistency of matter, co-emergent with the artist's un-self-conscious thought. This approach and practice can produce images of completely abstract 'landscapes' (or dreamscapes), gravel and moss, erosion, or constellations, that give rise to rich associations in our reverie. Dubuffet called such works 'landscapes of the brain' that open onto the 'inner dance' of the painter's mind (Dubuffet et al. 2006, 119).

Diverse artists in the modern era have pursued this goal. In Gutai, the Japanese artistic movement, their manifesto of 1956 clearly outlines a philosophical and aesthetic attitude to matter that is shared by many artistic practices discussed in this book:

[P]aint, pieces of cloth, metals, clay or marble are loaded with false significance, so that, instead of just presenting their own material self, they take on the appearance of something else. Under the cloak of an intellectual aim, the materials have been completely murdered and can no longer speak to us. Lock these corpses into their tombs. Gutai art does not change the material: it brings it to life. Gutai art does not falsify the material .... The spirit does not force the material into submission. If one leaves the material as it is, presenting it just as material, then it starts to tell us something and speaks with a mighty voice (Yoshihara [1956] 1996, 695).

Similarly, Carl Andre explains that he is 'less interested in imposing a form on matter than revealing the properties of the matter' (quoted in Cummings 1979, 187). Robert Morris, referring to a stick that allows paint to drip rather than being controlled by a brush, writes that it is 'in far greater sympathy with matter because it acknowledges the inherent tendencies and properties of matter' (Morris 1993, 43). He adds:

Certain art is now using as its beginning and as its means, stuff, substances in many states – from chunks, to particles, to slime, to whatever – and prethought images are neither necessary nor possible. Alongside this approach is chance, contingency, indeterminacy – in short, the entire area of process (Morris 1993, 67).

Receptivity and attunement to unorganised matter in artworks, celebrating contingency and process, became the stated aim of a number of postwar artistic practices, producing new ways of making art. It also led to a shift towards a psychology of viewing art that had hitherto been only implicit or peripheral. These observations lend new perspectives to abstract art, at least as it has been theorised, which I discuss further in Part III.

It is important to recognise that both the production and reception of a work of art are activities that share a 'non-cognitive' element. It is interesting that in the realm of reception this 'abandoned' matter barely makes an appearance when art history is written, perhaps because such writing is always already a project of meaning construction and organisation. Krauss is an important critic of the pure opticality of high modernism and its pretensions to being 'the display of reason, of the rationalized, the coded, the abstracted, the law' (Krauss 1993, 21), which she believes leads to an increasingly positivist art history. An important corrective to the cognitive understanding of vision, according to which modernist art requires conscious and willed looking, is that this kind of vision, in convoluted ways, represses the 'optical unconscious', the influence of barely conscious desires, fears and drives when looking at art. On this view, looking at matter painting that references or uses earth and soil, dirt and the scatological, only to extract semiotic or cognitive information is a form of repression.[12]

We are not just deciphering images but also registering the rhythms they suggest, which become internalised as cues for wordless thought.

[12] According to Mauzerall, even Krauss's acute observation of the material forces in artworks is part of a mechanical Freudian psychoanalytical system of interpretation, which also 'reduced matter to a sign' (Mauzerall 1998, 92). This can be seen in Krauss's interpretation of Max Ernst's *The Master's Bedroom* (1920), where she suggests that the artist consciously structured his painting to illustrate various Freudian theories about the mechanics of repression.

Sylvester points to a similar, longstanding distinction between saying something, using linguistics or semiotics, and showing something, which 'corresponds to Wittgenstein's distinction between what can be *said* and what *shows* itself, and the point about art is that it shows rather than says' (Sylvester 1997, 465).[13] What 'shows itself' is felt, intuited, and seems to come and go in a rhythm that is very different from the mechanical seriality of the everyday, which we are conditioned to internalise:

Ideas are inscribed *in* actions whose repetitive, ritualised performances are borne by concrete individuals who are thereby practically constituted as compliant or agentic subjects. While such performances are institutionalised in rituals and ceremonies, they also become sedimented at a corporeal level, where they are repeated as habits or taken for granted know-how: lodged in the bodily memory that Bourdieu calls habitus or phenomenologists refer to as a lifeworld. It is indeed this nonreflexive habituality and the way it imbues objects with familiarity that makes artifacts, commodities, and practices seem so natural that they are not questioned. It is in this sense that ideology or power operate most effectively when embedded in the material, practical horizons and institutions of everyday life (Coole and Frost 2010, 34).

Importantly, this complex machinery runs on linear, repetitive rhythms and methods, anticipated and regulated by the disciplinary controls of the base and superstructures of capitalist economies and technologies. On this view, artworks can produce ruptures in the regular and rhythmic functioning of this machinery. As the philosopher Henri Lefebvre proposes in *Rhythmanalysis* (2004), the continual production of the everyday is based on rhythmic repeats: bodies, actions, processes, machinic automations and transport, the structure of work and play, interactions and communications, all attempting to create order and secure 'the normal', as they measure the psychological and somatic production and passage of time.[14]

---

[13] Michel Foucault might have added that the two are not commensurable or translatable, although we repeatedly try to make them so.

[14] This thesis by Lefebvre (2004) shares synergies with Hayles (2017), where Hayles pieces together the noncognitive unconscious routines of our everyday life that become the mode of existence that prevails over the exercise of conscious, political and ethical agency. What is not part of this argument is how art and cultural forms can resist the mindless drift into the noncognitive unconscious manipulated by economic and social systems by producing its own kind of noncognitive unconscious or semi-consciousness. It is my argument that art, particularly abstraction, produces a noncognitive unconscious without these manipulations to conform, to consume, to drift. The key is to exercise a cognitive unconscious that is creative rather than normative. For an interesting analysis of how technological and economic systems operate to regulate psychology and behaviour, see Crary (2013). One would think that events in a world dominated by the continual horror of violence, war and displacement would disrupt the serial repeats of learning by rote, shopping, working, watching television. Although these events appear to contradict the view that we are trapped in everyday serial repeats and perceptions, we

Historical Materialism

Is an automatist sketch by Masson self-indulgent and reactionary because it fails to raise consciousness about social inequality or environmental degradation? Some of Herbert Marcuse's philosophical writings, particularly his essay *The Aesthetic Dimension*, continue to be valid today as an important critique of this kind of reductionism. In crude Marxism, materialism would require that the social relations of production must be reflected in the artwork as 'part of its inner logic and the logic of the material' (Marcuse 1978, 2). He notes:

This schema implies a normative notion of the material base as the true reality and a political devaluation of nonmaterial forces particularly of the individual consciousness and subconscious and their political function. This function can be either regressive or emancipatory. In both cases, it can become a material force (Marcuse 1978, 3).

This is ignored or passed over in vulgar materialism, where consciousness and the unconscious, emotions and the imagination, are 'dissolved into class consciousness' (Marcuse 1978, 3).

For Karl Marx, this leads to an 'abstract materialism' that elevates or intellectualises matter in order to absorb it effectively into mind. In *Critique of Hegel's 'Philosophy of Right'* (1844), Marx attempts to show how this kind of abstract materialism is linked to an ideological system that produces the subjugation of brute matter, understood as chaotic, in a hierarchical organisation of society. Marx castigates Hegel for positing the extremes of matter and mind, and then trying to reconcile them. He writes that, for Hegel, 'each extreme is its other extreme. Abstract spiritualism is abstract materialism; abstract materialism is the abstract spiritualism of matter' (Marx [1844] 1977, 89). According to Marx, Hegel could achieve this reconciliation of two opposing abstractions precisely because they were abstractions. The concept of 'mind', according to Marx, is not in opposition to 'real' matter: both are abstractions from matter that actually exists. He associates Hegel's abstractions with idealism. Marx detects an analogy between Hegel's natural philosophy, which rejects atomistic theory (that the cosmos is chaotic and made of a multitude of atoms), and Hegel's philosophy of right. This is because for Hegel, and some neo-Hegelians, the ordered system of the state develops from a unified and ordered cosmos, and is thus justified. On this view, atomistic theory would undermine the unity of law, divine order, nature, the ethical world, rational process and an ordered reality.

---

[14] should ask whether the everyday is in fact a kind of narcotic that makes such ruptures seem remote.

Although matter, and particularly matter in motion, is for Marx a fundamental substrate of thought, which he sets against Hegelian idealism, he also criticises advocates of materialism who ignore the importance of the social and historical reality by which matter is transformed into specific practices and systems. For Marx, although matter is primary, it is made knowable by how it is used, exploited or hidden by a system of human interactions – matter is thus known through practice within specific historical contexts. Because this is so, it is possible for humans in the realm of practice and social exchange to transform matter. Whether this leads to a reinvestment in the kind of hylomorphism that values form and rational will over matter, subjugating matter, is a moot point.[15]

According to Marcuse, Marxist aesthetics is in danger of devaluing the inwardness of sensations, emotions and non-conscious processes, which are in themselves material forces, because they do not illustrate or map clearly onto class ideologies. This scorn for inwardness 'is not too remote from the scorn of the capitalists for an unprofitable dimension of life' (Marcuse 1978, 38). Rather than being irrelevant to the class struggle because it does not directly articulate oppression using the language of reason, such inwardness may act as a place of refuge, a bulwark against the ubiquity of the ideologies of capitalism, consumerism, authority and statehood. These ideologies infiltrate every aspect of human life, demanding that our time and inner life be accounted for and made purposeful using the correct methods of rationalistic cognition. The value of Marxist aesthetics is to remind us that our experiments with artistic thought and creativity, and our ability to spend time on them, stem from privileges of class, education, wealth and power.

Marcuse believes that art is revolutionary because it can be 'subversive of perception and understanding, an indictment of the established reality' (xi). This established reality is secured by rational hierarchies of the use value of the products of creativity, limiting chance and non-rational processes that allow for different ways of understanding hierarchies. Art's power consists in fulfilling a cognitive function: 'it communicates truths not communicable in any other language; it contradicts' (10). But it contradicts in ways that negate the reality principle, formulaic narratives, the teleological and positivist faith of science, institutions and the individual ego in whose name neoliberalism is allowed its worst excesses. As Malcolm Miles notes, Marcuse's argument is that:

---

[15] The artist Carl Andre notes: 'Matter as matter rather than matter as symbol is a conscious political position I think, essentially Marxist' (quoted in Siegel 1970, 178).

art interrupts codes and structures (as of perception) which affirm the dominant order of society, [and] especially when read as the beautiful, [art] is radically other to a world of oppression and repression, so that its presence fractures that realm's surfaces. Art, then, is for Marcuse the location of a radical force against the unfreedom of life under capitalism or totalitarianism (Miles 2014, 12).

When Marcuse writes, '[i]f people were free then art would be the form and expression of their freedom' (Marcuse 1978, 38), this freedom would undoubtedly consist of an escape from the prescriptions of a rational order that suppresses spontaneity. According to Daniel Belgrad, for the American poet Charles Olsen spontaneous composition was

the key to poetry's social relevance. . . . [H]e believed that the conscious mind was the gatekeeper of social proprieties; social alternatives were therefore available first only at the unconscious level. Spontaneous composition avoided the falsifications introduced by a conscious mind that internalized ideological standards. By offering unmediated access to unconscious thought processes, spontaneity provided a vantage point from which to question the culture's authority and created the potential for authentic communications exploring new forms of human relatedness (Belgrad 1998, 29).

Belgrad argues that artists of the 1950s valued acts of spontaneity as a form of freedom, resisting the tight controls and structures imposed by conservative America over social, cultural and economic formations. The artists and writers who were part of this 'culture of spontaneity' did not have a systematic and coherent programme but were united in trying to escape formula, dogma and the kind of control associated with conservative authority and academia. Belgrad writes that 'the practice of spontaneous improvisation was released by "psychic autonomism" adopted by American artists and writers at the outset of World War II' (Belgrad 1998, 9). This was a technique of writing or drawing in an absentminded or spontaneous manner, important for 'bringing ideologically inadmissible possibilities into awareness' (9). Although the products of this spontaneity were hard to identify with political action, they could be seen as resisting respectable conservative ideals that demanded art be worthy, didactic and, above all, ordered.

There is one sense in which the Surrealist attempt to open up the possibilities of artistic practice beyond reason shares synergies with American abstraction. Belgrad gives an account of the culture of spontaneity, involving the beat poets, abstract expressionists and jazz musicians of the 1940s and '50s, who were inspired to escape the inhibitions of bourgeois culture and authority, some championing 'primitive' or ancient art, which they believed escaped the straitjacket of reason. Although these heterogeneous responses were in many cases essentialist and exoticising,

they were used to attack American conservatism and academic institutions sympathetic to scientific positivism, which cooperated with the aims
of the industrial-military complex. Reduced to a stark opposition, 'the
primitive', for those invested in the culture of spontaneity, meant access
to the non-rational world of dreams and myths, and the somatic dimension through which this world could be experienced. For artists, it also
involved modelling matter into art. As Belgrad writes, 'the abstract
expressionists did not try to depict Native American life, but to use the
forms of Native American art as a "medium of thought"' – or, as the artist
Adolph Gottlieb put it, the 'free association of images and symbols which
I couldn't explain' (Belgrad 1998, 63). The composition, the rhythms
and intervals, and the affective aspects of the symbols and textures of
Native American art were used as a way of thinking beyond cognitivist
approaches and syntactical and semiotic decoding. By arranging space,
rhythm, colour and plane, gesture and composition, and allowing the play
of sensation, chance, intuition and automatism, these artists escaped the
premeditated and programmatic rational uses of their art, which would
turn their works into didactic illustration. Similarly, by learning more
about context, cultural comparison and art history, one can sometimes
become more attuned to these rhythmic and felt aspects of artworks.[16]

For Belgrad, the culture of spontaneity was part and parcel of the move
away from thinking about matter, the universe and art in ways that
stemmed from centuries of inferences from Euclidean, Newtonian and
Kantian logic. Although primitivism was encrusted with layers of exoticism
or mysticism, it provided artists with new ideas concerning composition
that avoided painterly conventions learned through rational means at

---

[16] Historically, there has been a tendency to explain abstract art either in 'contextualist'
terms, as a superstructure arising from economic, social and political circumstances, or
with a 'formalist' emphasis that is more ahistorical. See Schapiro (1937) for a well-known
attempt to suggest the historically specific social and economic paradigms underlying
innovations in art such as abstraction. Cheetham (2006) argues persuasively that formalist purity and autonomy, and the 'infection' of social relevance, have always struggled
with each other in the history of abstract art. Halley believes that 'abstract art is simply the
reality of the abstract world' (1997, 32), a massive network of geometrically organised
urban spaces, cells, circuits, apartments and highways that create a spatial and rhythmic
sensibility (or alienation) that find its way into the psychology of design and the making of
abstract art. Works on abstract art that I engage with in Part III are sensitive to historical
conditions but not in ways that are easily understood as commensurate. Similar points are
made by Varnedoe (2006); Schimmel (2012) (this is interesting in terms of abstraction
which is not geometrical, such as *Informel*); Rosenthal (1996); Belgrad (1998) and others.
It is important to acknowledge how historical events are internalised by psychological
processes to be externalised back into history through artistic practices. For
a psychoanalytical approach to abstraction, which I disambiguate from a psychological
approach in Part III, see Fer (1997). Houston attempts to ground 'perceptual abstraction', such as Op Art, 'devoid of narrative and symbolic content in its historical conditions' (Houston 2007, 57).

schools and academies, and was therefore closer to their own dream worlds and reveries. For Bachelard, reverie is a creative mental process that fragments the composition of rational and ideal forms. These fragments are irrational images and juxtapositions that we see in Dada collage – myth, contradictions, paradoxes and metaphors that give rise to a 'decomposition of forces', which breaks with 'the naïve and egotistical ideal of the unity of composition' (Bachelard [1948] 2002, 111).

For the critic Harold Rosenberg, Willem de Kooning's spontaneity was political particularly because it was 'repugnant to bureaucrats, conformists, organization-men, and programmers' (De Kooning and Rosenberg 1974, 36). In 1958, when many of the post-war artists in Europe and America who I examine in this book were producing their most experimental abstract works, Rollo May articulated a view of existentialism as a way to break free from the 'Victorian man' and his rational compartmentalisations, which many artists were intuitively struggling against:

The Victorian man saw himself as segmented into reason, will, and emotions and found the picture good. His reason was supposed to tell him what to do, then voluntaristic will was supposed to give him the means to do it, and emotions – well, emotions could best be channeled into a compulsive business drive and rigidly structuralized in Victorian mores; and the emotions which would really have upset the formal segmentation, such as sex and hostility, were to be staunchly repressed or let out only in orgies of patriotism or on well-contained week-end 'binges' in Bohemia in order that one might, like a steam engine which has let off surplus pressure, work more effectively on returning to his desk Monday morning. Naturally, this kind of man had to put great stress on 'rationality.' Indeed, the very term 'irrational' means a thing not to be spoken of or thought of; and Victorian man's repressing, or compartmentalizing, what was not to be thought of, was a precondition for the apparent stability of the culture. This compartmentalization went hand in hand with the developing industrialism, as both cause and effect. A man who can keep the different segments of his life entirely separated, who can punch the clock every day at exactly the same moment, whose actions are always predictable, who is never troubled by irrational urges or poetic visions, who indeed can manipulate himself the same way he would the machine whose levers he pulls, is of course the most profitable worker not only on the assembly line but even on many of the higher levels of production (May 1958, 21).[17]

[17] It is interesting that this description of organising daily routines into highly organised spatial cells and compartments and temporal intervals has been likened to the capitalist system of Taylorisation, a geometrical urban, consumer and digital network that is itself abstract. Indeed, abstract artist Peter Halley has tried to paint this inescapable geometric maze in which we are all physically and psychologically immersed. He writes:

[P]eople live in sealed houses or condos in highly controlled landscapes. They travel in the sealed environment of the automobile along the abstract pathway of the highway to equally artificial office parks and shopping malls. When one speaks of abstract art, it is essential to remember that it is only a reflection of a physical environment that has also become essentially abstract (Halley 1991, 60).

It is interesting that this view of the Victorian man, and the underlying segmentation and rhythms of life, are still valid today. But in all these examples, it is art that was seen to offer a way to exist parallel with reason. Even much later, Deleuze and Guattari see art as being able to produce sensations and affects that undermine not only the Victorian man and the bureaucratic mindset but also the kind of citizen, ego and normative subjectivity that may be seen to flow from it. In *Francis Bacon: The Logic of Sensation* (2003), Deleuze discusses rhythms, vibrations and sensations which overflow sense, decorum and dogma, sharing important synergies with Lefebvre's 'rhythmanalysis', which, as I have mentioned, is also a critique of the segments and rhythms produced by the rationalisation of modern life.[18] It is with Bachelard, however, that this rhythmanalysis is linked to the traces of oneiric memory in present perception which artists and writers are able to model

Chaos

For Deleuze and Guattari, in intensive encounters with art, the brain plunges into chaos (1994, 215) and this could be what happens with some powerful artworks that move the viewer. Deleuze writes about Francis Bacon's 'defacement' of his subjects, brought about by polychromatic strokes and thick impasto, which appeals to direct somatic engagement rather than detached intellectual examination. Nevertheless, as Deleuze and Guattari note, art 'is not chaos but a composition of chaos that yields the vision or sensation, so that it constitutes, as Joyce says, a chaosmos, a composed chaos – neither foreseen nor preconceived' (Deleuze and Guattari 1994, 204).

Scientists have studied in detail the chaotic dynamics of the brain, and embodied emotions and their dynamics. Notable in this regard are Walter J. Freeman, György Buzsáki, Ichiro Tsuda, Emmanuelle Tognoli and J. A. Scott Kelso, who present strong evidence that the brain operates with nonlinear dynamics, metastability and chaotic itinerancy, providing empirical support for the idea of 'chaos in the brain', understood not as stochastic but as an interaction between order and chaos. Many of these researchers have observed dynamic changes in brain activity between different regions and networks, which operate in rhythmic and syn-chronised ways. Yet the depth of research in this area rarely, if ever, extends to the behaviour of individuals engaged in viewing and

---

[18] This supports Jones's (2005) analysis of the bureaucratisation of the senses. Jones studies how Greenberg's emphasis on opticality, or the disembodied eye, was the product of a positivist ideology that insisted on the superiority of the detached intellect while denigrating emotional or affective entanglements with art.

making art. In neuropsychology, the study of art is dominated by the reductionist paradigm of examining specific brain areas rather than the dynamic complexity of their interactions.[19] Although nonlinear dynamics and theories of the brain inform some of the language and ideas of Deleuze and Guattari, and those influenced by their work, these ideas would benefit from the precision provided by empirical studies on 'chaos in the brain', which I attempt in Part II. But it is interesting as well that these more recent scientific studies often support the general statements and speculations about the functioning of the brain in Deleuze and Guattari.[20]

N. Katherine Hayles writes that in chaos theory

regularities of the system emerge not from knowing about individual units but from understanding correspondences across scales of different lengths. It is a systemic approach, emphasizing overall symmetries and the complex inter-actions between microscale and macroscale levels (Hayles 1990, 170).

This approach requires an ability to cross-reference different levels of description. Hayles writes that, for many scientists

chaos is more than just another theory. It represents an opening of the self to the messiness of life, to all the chaotic unpredictable phenomena that linear science taught these scientists to screen out. Once roused, they remember that the messiness was always there. Moreover, now they are able to see nonlinearity in a new light, perceiving it as central rather than marginal, beautiful rather than aberrant (170).

---

[19] There are too many examples to mention but Zeki (1999) and Livingstone (2002) are well-known scientists who have developed theories about our involvement with art from a series of experiments on brain cells, neurons and particular brain areas responsible for processing the basic features of art, such as line, colour and luminance, based on data from fMRI scans, lesion studies and optical illusion diagrams. These approaches look at basic perceptual responses to abstract art and emphasise formalism as triggers for shape and motion detection in geometric forms, colour and luminosity. A good summary of these approaches can be found in Weir and Mandes (2017).

[20] A good example is Grosz (2008), which centres on Deleuze's understanding of sensation in art resisting overly semiotic and mechanical approaches: sensation 'extracts' from chaos (Grosz 2008, 61), art 'protects us from chaos' (56), or wards off chaos. Sensations pass through the body. No particular biological details of how sensations do this or are transformed by the body are examined. Deleuze and Deleuzians are fond of Bacon's notion that art should directly have an impact on the stomach rather than the brain, that it could have a direct effect on the nervous system. It would help to drill down into the details of how this might actually work in terms of neuronal assemblages, oscillations, neurotransmitters, sensorimotor and visceroceptive and other metabolic systems. But this is rarely done, and so the expression often languishes at the vague and general level, even though Deleuze is at pains to go into a lengthy rhythmanalysis of Bacon, exploring a large number of different ways of describing vibration and sensation. Similarly, in Grosz there are some very interesting passages where she describes vibrations and rhythms, and offers various explanations for how sensations are composite or have layers or levels (83).

In the Western tradition, chaos has played the role of the other – the unrepresented, the unarticulated, the unformed, the unthought. Hayles's understanding of chaos is that it is not only a fundamental property described by science but also a method of understanding this property. This 'way of thinking' is important for the humanities as well: 'both scientific and literary discourses are being distinctively shaped by a reevaluation of chaos. It is this vision that defines the contemporary episteme and differentiates it from the modernist era' (Hayles 1990, 177). My study is a product of this contemporary episteme.

While it is my contention that it is important to understand art, literature and many other domains of human activity in terms of the interplay of chaos and order, there have been various critics of such attempts. Stephen H. Kellert, in *Borrowed Knowledge* (2009), provides a detailed study of the controversies in the literature, and evaluates both the work that employs such methods and its critics. Kellert draws on the work of various scientists who attacked French intellectuals such as Jean-François Lyotard and Michel Serres. As Kellert points out, these scientists were not opposed to the idea of extrapolating concepts from one field to another, but only against extrapolations that are unsupported. Some of the main misunderstandings and unsupported extrapolations are as follows:

1. Enthusiasm for chaos theory has led some writers to overstate its importance, claiming that chaos is a fundamental 'truth' of the universe. But not everything is chaotic.
2. Kellert is at pains to show that sensitive dependence (the 'butterfly effect') has been widely misunderstood and 'should not be conceived of as a story about the special causal powers of one specific butterfly; in a chaotic system, each and every butterfly matters' (Kellert 2009, 9).
3. Kellert explains that a certain pattern, regularity or consistency arises after observation. Chaotic routines may eventually exhibit patterns called 'strange attractors', where the duration of movements and changes of the 'chaotic' system seem to bottom out or gravitate towards a more predictable trajectory as the system becomes more stable or reverts back to learned patterns, drives, memories or stable reference points in the environment. Thus these 'strange attractors' are not 'mysteriously' magnetic.
4. Writers have confused true stochastic behaviour with chaos theory, which is nonlinear behaviour but is still deterministic behaviour:

[C]haos theory is the study of unpredictable behavior in simple, bounded, deterministic systems. Such behavior is extremely complicated because it never repeats, and it is unpredictable because of its celebrated sensitive dependence on initial conditions: even extremely small amounts of vagueness in specifying where the system starts render one utterly unable to predict where the system will end up ... chaotic behavior is investigated using the mathematical tools that go by the name 'dynamical systems theory' or just 'dynamics' (Kellert 2009, 6).

5. Chaos theory in science and mathematics is quantitative and exacting. It is important to note that this kind of analysis is not the kind that is employed directly in the interpretation of artworks.[21]
6. The unpredictability of chaos is not the same as indeterminacy or Heisenberg's uncertainty. It should also not be confused with complexity theory, 'which studies the intricate behavior that can result from large numbers of subsystems interacting according to simple rules' (Kellert 2009, 8).

In his conclusion, which I embrace, Kellert is broadly positive about 'borrowed knowledge', offering the following attempt to achieve a balance:

[I]t is indeed important to 'get the science right,' since a mistaken use of a scientific theory or term may disrupt the effectiveness of a metaphorical borrowing. But at the same time, the mathematical structure of a theory does not fix its broader cultural meaning; a narrow focus on scientific correctness ... runs the risk of missing out on the very real meanings attached to scientific results ... any attempt to portray such cross-disciplinary transfers as simply misunderstandings of the science rules out the possibility of exploring the impact that science can have on the practices of other disciplines (Kellert 2009, 233–234).

I am also buoyed by the fact that Kellert approves the use of aspects of chaos theory in the work of Walter J. Freeman, 'who demonstrates the possibility of fruitful applications of the techniques of nonlinear dynamics within neurophysiology' (Kellert 2009, 7), which is what I attempt to do in Part II. Kellert's study explains my reluctance to adopt chaos theory, tout suite; I prefer the idea of 'metastability', which can be understood as a (nonlinear) dynamic complexity which features aspects of order and disorder.

The overall functioning of the brain has been described as 'metastable', where metastability, a term taken from nonlinear dynamics, is understood as a process in which ordered set routines and improvisation emerge in different parts of the brain, interacting with each other, in response to environmental stimuli. Bruineberg and Rietveld write: 'we need to understand how the self-organized metastability of the brain-body-environment

---

[21] But the conclusions reached by these studies are extrapolated and theorised as part of a theory of the psychology of art in Part III, particularly when these scientific works use examples of 'external perturbations' and other similar examples to construct brain-body-world looped interactions and metastable coordination.

system interacts with the self-organized metastability of the brain' (Bruineberg and Rietveld 2014, 10). This understanding of the brain as spontaneously organised and continuously pulsating with rhythms and oscillations that gather and disperse in interaction with the world has made negligible impact on neuroaesthetics or neuroscientific and psychological studies of art, with the result that we are left with rather more 'static' studies of brain states and artworks. Self-organisation is a spontaneous process where global order, rhythms, synchronies or other patterns of coordination emerge out of local interactions between components of an initially disordered system. It can be observed across a broad range of biological, chemical and cognitive phenomena: 'Common examples include crystallization, the emergence of convection patterns in a liquid that is heated from below, chemical oscillators, swarming in groups of animals, and the way that neural networks learn to recognize complex patterns' (Marabissi and Fantacci 2015, 13).

In neuroscience, metastability describes how neural populations work by negotiating aspects of stability and instability, self-organising into larger rhythmic and arrhythmic movements. What at first seems arbitrary or arrhythmic may in fact be a complex set of rhythmic interactions over time. The neuroscientist J. A. Scott Kelso explains that the 'perceptual system (and the brain itself) is intrinsically metastable, living at the edge of instability where it can switch spontaneously among collective states' (Kelso 1995, 223). The challenge for those who seek a deeper understanding of how art affects us is to understand that this metastability emerges not just in the brain but in cooperation with other aspects of the body (the nervous system, muscle groups) that are engaged in the process of viewing or experiencing art. Although precise in its methods and calculations, even nonlinear dynamics would buckle under the strain of dealing with so many variables, but it is important that we open our minds to the idea of a complex metastability, composed of both order and chaos, that is sustained across brain, body and artwork.

This is particularly the case in the process of painting. In *Brushstroke and Emergence*, James D. Herbert suggests that the process of painting is a 'unified mental/material continuum not clearly subdivided into differentiated components' (Herbert 2015, 18). This unified continuum is not an ordered one that proceeds in linear steps. Gerald Cupchik suggests that such art-making can be seen as a 'recursive and stochastic process' where attention to the emerging composition switches from part to whole and back again in a 'motor-sensory loop' (Cupchik 2016, 271). This presents problems for reductionism, that is, the breaking up of a complex set of entangled events in order to study one or two isolated snapshots which are assumed to explicate the whole event. Similarly, the idea of emergence is about larger events that cannot be traced back to their origins in a linear causal chain. As dynamic systems theory shows, the collective action of various elements can create something irreducible, unpredictable and novel. Related to this idea is the work of philosopher Alva Nöe, who points to larger events, the 'things themselves', rather than the neurons processing aspects of experience:

What is the causal substrate of the experience of the wine's flavor? Perhaps this substrate is only neural, but perhaps it is not. Perhaps the only way to produce flavour sensations is by rolling a liquid across one's tongue. In that case, the liquid, the tongue, and the rolling action would be part of the physical substrate for the experience's occurrence (Nöe 2004, 220).

This holistic approach challenges the 'computational' understanding of the mind, which may be reductionist in assuming that higher-order phenomena (such as tasting wine) can be:

collapsed into lower order ones and that lower order phenomena were the ultimate explanatory elements, the 'causes' that science [seeks] to isolate. Yet the breakthrough of nonlinear dynamics has shown us that explanations of self-organizing phenomena can only be given in terms of the qualitative forms of behavior of the system as a whole, i.e., in terms of system properties that resist analysis in terms of the properties of the parts, whether they be individual neurons or discrete input to the system. This implies that in explanations of self-organizing brain dynamics, there necessarily will be relative independence from the nature and properties of the substrate; hence micro-reduction, the aim of traditional explanations, does not work (Freeman and Skarda 1990, 281).

What is required is to challenge the privileged position given to reductionism in the interpretation of art, an approach that is supposed to penetrate, reveal and extract meaning from artworks by breaking down their complexity into 'chunks' of perception. This privileging also assumes that the human factor is the most important in a network of factors involved in the production (and experience) of art.

The rhythms and textures of matter suggested in abstract art, which lie outside of or beneath the exercise of conscious, formal logic, help to produce in the viewer less-than-conscious and more arbitrary mind wandering and reverie. This opens up a new way of understanding art, and of looking at and understanding how we (our brains) respond to art. It is this subliminal and rhythmic process that I propose can be studied by bringing together various levels of description. Each discipline has a different contribution to make to our understanding of rhythm and how it emerges uniquely in each domain of brain, body and world. This book attempts to show how the traditional duality of mind and matter is better understood as differences in degree.

My argument is that the activity of making and viewing abstract art, particularly matter painting, hovers around the mid-point in this scale.

In my last book, *The Psychology of Contemporary Art*, I explored the cognitive and conceptual complexities that modern and contemporary artworks help to produce. The present volume complements the earlier work by examining how rhythm and matter in art produce sensations, feelings and mind wandering. Both philosophy and psychology, and the latest research on rhythmic oscillations in the brain, suggest that non-conscious and daydreaming states are pulsative, and can cooperate with the rhythmic patterns that artworks suggest.

Each discipline has a different contribution to make to our understanding of rhythm and how it emerges uniquely in each domain of brain, body and world. Part I analyses terms such as mind, matter, rhythm and sensation informed by new materialist philosophy. Part II is a meta-analysis of current research in neuropsychology on brain oscillations, and the phenomenon of mind wandering, which abstract art helps to sustain. And Part III demonstrates how these different disciplines can interact in case studies of specific works of art which help to cue rhythmic sensations.

Part I
Rhythms of Mind and Matter

1.1   Philosophical Approaches

For Aristotle in the *Physics*, change is a fundamental principle. For change to take place, a subject must lose or gain form. Aristotle uses the example of bronze, a formless material, which gains form from the cast to become a statue of Hermes, the messenger of the gods, and the god of boundaries and trespass. The thing that undergoes formation or change in Aristotelian physics became known in Scholasticism as *materia prima*, 'primary matter'. For matter, Aristotle used the Greek word *hylē*, which literally means a forest or woodland. The term also refers to timber or 'wood cut down', and in a further derivation it refers to the timber used in the construction of something (Leclerc 2004, 115). *Hylē* is a general term for the thing from which something else comes, such as the bronze of a statue or the marble of a human figure. In Christian texts, notably in Thomas Aquinas, *hylē* is understood as dumb and passive, 'receiving' the *eidos* (form), whereas form is thought of as the more important active force that creates a boundary around the amorphousness of matter. But Aristotle also suggests that form and matter struggle with each other (McCumber 2013, 101). Form is assumed to be immaterial, mental and ideal, while matter is none of these things.

For the philosopher René Descartes in the seventeenth century, matter is beneath us or outside us. This view characterises thought as mind over and above matter. The physicist Isaac Newton elaborated on Descartes's understanding of matter by identifying its 'primary qualities', such as hardness, impenetrability, mobility and inertia, while relegating colour and taste as secondary qualities. With these theories, the distinction between external reality and internal categories becomes entrenched.

A century later, the philosopher Immanuel Kant presented an ordered system of mental states, including sensations. He writes that 'sensations are ordered in sequences, one after each other in time, and one next to each other in space, following the laws of geometry and mathematics' (Wicks 2013, 53). What seems important is the direction of Kant's thinking, in which cognitive categories are applied to the contingent. In his *Critique of Pure Reason* (1781), contingency and chaos are not positive attributes, and it is assumed that their only value is as 'raw material' to be worked on by the logical 'manifolds' (branches in the structure of the mind). As Robert Wicks explains, for Kant there is a formal structure that underlies our fundamental sense of self, the transcendental unity of consciousness: 'at the core, this unity of consciousness is the source of all stability and synthesis' (Wicks 2013, 84). According to Paul Churchland, for Kant, 'things-as-they-are-in-themselves (independent of human perception and conceptualisation) are forever unknowable by humans' (Churchland 2013, 138). On this view, matter must also be unknowable, particularly if it is uncountable and vast, like the churning of the sea or the vault of the starry sky which sends the measuring mind into overload, a condition described as the sublime.

Referring to Kant, the philosopher Jean-François Lyotard writes that the sublime is 'what dismantles consciousness, what deposes consciousness, it is what consciousness cannot formulate' (Lyotard 1991, 90). For Kant, however, beholding the boundless sea or starry sky provides an opportunity to consider the distinction between the beautiful and the sublime. The philosophical and psychological properties associated with beauty tend towards form; they are composed by and appeal to the contemplative mind, whereas the sublime is formless and disintegrated, agitating both the mind and the body (extended into space) as it loses itself in nature (or matter).[1] For Kant, the sublime involves the mathematical and the dynamical, and, in extreme cases, when it is immeasurable, it can be terrifying. Ultimately, however, for Kant 'the boundlessness of natural objects, such as the raging sea or the starry sky, leads us only to the appreciation of what is truly sublime and "powerful", which is nothing else but reason' (McCumber 2013, 119).

---

[1] Kant discusses the free play of the imagination, which synthesises sense-perceptions but is nevertheless bound to rules of procedure, referring to this as 'free lawfulness' or 'lawfulness without a law' (Ginsborg 1997, 37). This idea is not clearly explained and has generated much debate. For a discussion of how this might relate to art, see Bell (1987). Bell suggests a painting by Pollock can be about

discovering in the diversity of sensory experience a felt unity, coherence, or order, which is non-cognitive and non-conceptual ... to enjoy a spontaneous, criterialess, disinterested, presumptively universal, non-cognitive, reflective feeling that certain diverse elements of experience as such belong together, that they comprise an intrinsically satisfying whole in virtue of their seeming to have a point (though without it being the case that there is some specific point which they are judged to have) (Bell 1987, 239).

The way in which we reason, using focused, voluntary thought, is elevated by Kant repeatedly. But often, reason does not play a leading role in our interaction with the natural world or in our experience of artworks. In fact, an excess of reason would interfere with the subtleties of experience offered by artworks, including Surrealist works and 'matter painting'. Surrealists, for example, dallied with the uncanny and ineluctable qualities of dreamscapes, where cognitive states or branches are not distinct in the manner suggested by Kant. Artworks are often produced precisely not to cooperate with the processing of the perceptual apparatus in an efficient manner, or to deliberately cause the viewer to stumble in awe of sublimated matter. The artist Jean Dubuffet suggested that the 'European thinker' overvalued reason, convincing himself that 'the shape of the world is the same as his reason . . . . I must declare I have a great interest in madness, and I am convinced art has much to do with madness' (Dubuffet et al. 2006, 115). This complements Surrealist attempts to link words, images, textures, processes such as collage and frottage, and chance methods in non-logical sequences to arrive at striking associations and dreamlike sensations. These could of course be described as nonsensical or 'mad', at least by those fond of using such labels. Dawn Ades writes that these techniques were aimed at producing mental activity outside conscious control, which

produced its own strange poetry . . . . Suspension of conscious mental direction was required for the practice of automatic writing but this was not dissimilar from the waking dream, or half sleep which was to produce some cherished images, just as dreams themselves were to do (Ades 1994, 13).

On the surface, viewed literally or from the perspective of the rational mind, such images and poetic juxtapositions may seem irrational and therefore worthless. One begins to 'see' that these artists were attempting to translate into tactile, spatial media dreamlike and thought-provoking sensations that are far from the neat and logical structures suggested by Kant.[2] These are not the sensations we normally experience during our waking state, determined by the exertions of cognition. Such art also appeals not to ordered sensations but to deliberately confusing multisensory experiences. This kind of unusual experience can lead to resetting automatic perceptions and produces a queasy, uncanny effect: a half-remembered dream or sensation experienced while awake, 'where dreaming and waking seem to slide into one another and which, turned outwards to the life of the streets . . . invite[s] bizarre encounters and experiences there' (Ades 1994,

---

[2] It is interesting that Heidegger writes: '[W]hen irrationalism, as the counterplay of rationalism, talks about the things to which rationalism is blind, it does so only with a squint (*schielend*)' (Heidegger 1967, 40).

13). This meandering state of being is sustained by attending to the broad outlines of an artwork for long enough before the rational, analytical self-consciousness returns to segregate such impressions and sensations from their deplorable intermingling. While Surrealist images evoke dreamlike rhythms and sensations and strange illogical associations, abstract art, devoid of figures and objects, invites us to lose ourselves in reverie. This does not mean that we abandon cognition and rational thought in order to experience 'irrational' art, but rather that we do not allow cognition to be the sole or most important factor in our encounter with art.

In fact, various forms of art go further than encouraging the alternation between rational control and abandonment to the senses. Art is often able to produce affective, chaotic and disorienting events. For example, there is the 'art of disturbation', as Arthur Danto termed it, such as we see with post-war art movements such as Vienna Actionism or Fluxus and kinds of performance art, where matter, the abject and the body are intertwined, aiming to return to the purely affective, anarchic and transformational roots of Greek theatre. This finds an ally in the poet Arthur Rimbaud's well-known objective, 'to reach the unknown by the derangement of all the senses' (Rimbaud [1871] 1966, 303–304). This is contrary to Kant's notion, so often used to regulate judgements about 'good' art, that 'space and time organise sensory information that is given to us, and since geometry and mathematics articulate the very structures of space and time, our experience must always have a geometrical and mathematical turn' (Wicks 2013, 59).

A philosophical counterpart of this swerve into chaos is Arthur Schopenhauer, in *The World as Will and Representation* (1818), for whom scientific logic and calm reason are limited and partial, made precarious by the deeper and more powerful desires and unconscious emotions that constitute the will. There are parallels to be drawn with the Freudian id, as the will is not a logical domain but subterranean and nonsensical, arbitrary and full of conflict. In a striking manner, Schopenhauer universalises this to the whole spatio-temporal world: everything is animated by this arbitrary force. This is echoed in various modernist artists' attempts to subvert the rational will in the artistic process, and in the work of musical composers striving to escape conventional forms with a direct appeal to sustaining subjective experience, and not necessarily in linear or sequential ways. Schopenhauer presents an alternative to the optimism of the vitalist tradition and forms of panpsychism that assume the universe has an underlying harmony and goodness. Influenced by Indian asceticism and Buddhism, the will, which fills the universe, was for Schopenhauer a source of misery. The only way that one could escape it was through asceticism or certain aesthetic experiences that are 'pure, will-less, timeless' (Schopenhauer [1818] 1909, 231). This is important as a way to foreground passive receptivity as a form of aesthetic participation. This idea bears some relation to the philosopher Martin Heidegger's *Gelassenheit* and applies to forms of artistic experimentation that require the will and rational cognition of the world to subside so that the earth can be disclosed, something that art enables and that is philosophically significant.

For Anton Ehrenzweig and many other art theorists, it was a long-standing aim to overcome the subject/object divide. Schopenhauer suggests that the separation between the perceiver and the perceived is abolished in aesthetic experience where the structures of reason, supporting judgements and inferences, temporal and spatial categories, and cause and effect are suspended: 'one can thus no longer separate the perceiver from the perception' (231). It is interesting that Schopenhauer speaks of matter revealing itself in the work of art, the qualities of light and stone in architecture (263), and of developing a sensitivity to materials disclosing internal forces, sentiments later to be found in Heidegger. Schopenhauer writes of water in fountains: 'we find it beautiful when it tumbles, rushes, and foams, or leaps into the air or falls in a cataract of spray' (282). And in gardening, it is not the will of the gardener so much as nature that reveals itself. But it is 'absolute' music, which puts aside all imitation of nature, cognition and narrative, that provides access to the thing-in-itself. Some of the descriptions that Schopenhauer reserved for fountains and music could be applied to twentieth-century abstract art's appeals to sensation, rhythm and the non-figurative. For Schopenhauer, there is something of us that we discover in art: it lifts the veil of cognition that divides us from things-in-themselves.

I mention Schopenhauer here not only as a contrast to the Kantian tradition but also because he believes feeling, subjectivity and chaos are forces that are not just to be subdued by order, knowledge and cognition but valuable in themselves. Kant aimed to examine the *a priori* conditions for making judgements about the properties of objects as the basis of aesthetic extrapolations: intuitions are guided by concepts. But Schopenhauer suggests that such conceptual contents – aesthetic judgements – interfere with direct aesthetic feeling. For Schopenhauer, it is 'intuitive cognition' (*Erkenntniß*) that we experience through art and nature, and one of the finest experiences this has to offer is a kind of will-less, timeless consciousness. Schopenhauer's various examples of the transition from the beautiful to the sublime, such as viewing an 'unbroken horizon, under a cloudless sky, trees and plants in the perfectly motionless air' (263), or the same scene 'stripped of vegetation, showing only naked rocks' (263), or the waves in a turbulent sea and craggy cliffs, generate feelings of pleasure or vulnerability (264). While opening oneself up to these intensities of intuitive cognition, one is being passively receptive to visual phenomena, entrained by their complex motion, rhythm and scale.

These moments of will-less experience could be compared to mind wandering and reverie experienced during meditation, while viewing art or listening to music, or while producing art. Although mind wandering involves autobiographical memory, where we imagine what was or what could have been, much of mind wandering is similar to daydreaming and, contrary to scientific accounts that make such episodes seem ordinary and reasonable, is full of all the marvellous and nonsensical things that can happen in dreams or in moments of tranquillity. Importantly, it was Heidegger who pointed out that the excesses of rationalism and mechanical reductionism we see in scientific method force phenomenal and sensory knowledge into the subjective realm. His philosophy challenges the polarisation between objective knowledge associated with scientific reductionism and subjective feeling considered to be the exclusive domain of aesthetics by proposing that art provides an opportunity for 'true existence' to emerge as a form of being-with that is greater than this artificial divide.

Rational categories and willed, voluntary thought are deferred during periods of incubation and creativity, giving free rein to involuntary associations. This will-less state is sustained by the neurodynamics of memory and involuntary thought, and requires some cultivation to be exercised in the appreciation of art. It involves an openness to chaos and a sustained period of metastability that defers the imperative to produce order and coherence from a constantly shifting visual field of abstract patterns. There are significant advantages to be derived from putting aside rational methods that often cannot see the *hylē* for the trees. Reverie involves a passive receptiveness to the world, a receptiveness that rhythmically flows with the disunity observed. Our own embodied rhythms are in synchrony with the rhythms of matter. This connection of internal and external worlds should be seen not as a reduction of movement-matter to Euclidean geometries or measures but rather as a deeper attunement of the body's rhythms to those in the environment so that thought itself seems to pulse with this connection. Thought does not proceed only from the mind, emanating outwards into the world through acts of measurement, reason, will and formal cognitive organisation. The world of matter, and particularly art, supplements, constrains and interferes with rational structures so that other looser structures emerge. Many artworks sustain this external-to-internal movement, helping to produce neural plasticity in unexpected, idiosyncratic ways.

A well-known study of jazz pianists by Limb and Braun (2008) found that, when musicians improvise and create spontaneous musical passages, the self-monitoring and planning areas of their brains, associated with the executive control network, are deactivated. Limb and Braun suggest that this deactivation 'may be associated with defocused, free-floating attention that permits spontaneous unplanned associations, and sudden insights or realizations ... to arrive at a solution without reasoning' (Limb and Braun 2008, 4). This deactivation is also 'responsible for altered states of consciousness such as hypnosis, meditation or even daydreaming' (5). They go so far as to explain the relevance of playing jazz to daydreaming: both require the vigilance of the executive control network to be deactivated, and both activities share 'defocused attention, an abundance of unplanned, irrational associations and apparent loss of volitional control' (5), features that are also associated with creative activity during wakefulness.

1.2   Henri Bergson

> For Bergson, perception is not a *synthesis* but an *ascesis*. Perception does not connect, it disconnects. It does not inform content but incises an order. It does not enrich matter, but on the contrary impoverishes it
> (Meillassoux 2007, 75).

Automatic perceptual routines disconnect us from matter in its astounding complexity and infinite variability, while sensations create relations with matter and can become part of its rhythms and movements. Perceptions are an efficient way to notate and extract from the chaos of matter shapes and forms that are relevant for navigating a world that is often designed to provide a spatial and semiotic formalisation of shapes and forms.[3] Perception is not a passive response to external stimuli, like a reflex, but actively parses and anticipates external events. Our tools and technologies, and the built environments that affect our everyday movements, are densely patterned sub-personal ecologies designed to be manipulated by (and products of) our perceptual routines. Seen another way, perception is a kind of well-oiled machine. For the philosopher Henri Bergson, 'our representation of matter is the measure of our possible action upon bodies: it results from the discarding of what has no interest for our needs, or more generally for our functions' (Bergson

---

[3] Kandel's understanding of the difference between sensation and matter involves quoting William James's opinion that perception recruits consciousness of further facts associated with the object causing the sensation (Kandel 2016, 111). The problem can be expressed in the following way: perception *supplements* sensation but in doing so it takes away from (extracts) the volatility or ongoing experiential complexity of the sensation qua sensation.

[1896] 2004, 30). A world in which this kind of narrow functional perception becomes dominant is disastrous for art practices that rely on the opposite principle: overriding perception. We can speak of the 'perception space' as something that is shared by the viewer and the artwork. While the viewer's experience of the artwork may include memories, concepts, feelings and reveries, the artwork will help to rhythmically cue or switch these kinds of mental states. An abstract painting can act as a 'map' to revisit these aspects of experience.

Bergson recognises that conscious perception 'shrinks' the world of matter, in the sense that it is highly selective and so closes off our awareness of much that is out there, but he pursues the idea that there is more to consciousness than this kind of contraction, that in fact consciousness can be relaxed ('detensed'), allowing an openness to the full and 'unedited' field of matter – matter without objects but with abundant and often arbitrary fluctuations. This suggests that the brain can be led or entrained by the activities of matter in a way that might not always extract what is useful or familiar or match what is being observed to preconceptions. One way of understanding this is to think of the brain that is mind wandering not as working to codify or measure the dynamics in the environment but, rather, as being occupied in low-level scanning, barely registering details and surrendering instead to an overall rhythmic impression, as is the case when one is listening to music, observing landscapes or feeling the grooves in the bark of a tree. We can be open to – or learn to be open to – densely interwoven multisensory rhythms in the environment to such an extent that we are not constantly subjecting each moment to cognitive analysis.[4] We do this often enough when the mind is at rest, when we are zoning out. Is it possible to acquire a sensitivity to abstract art and matter painting in the same way?

---

[4] Jones (2005) argues that Clement Greenberg's criticism in the 1950s and aspects of abstract art were typical products of a world where the bureaucratisation and separation of the senses were the order of the day. It is well known that Greenberg favoured pure opticality over the other senses, thereby demonstrating a form of self-mastery and detachment. The wide circulation of these ideas does seem to suggest a historically specific visual regime. However, Pollock's titles for paintings executed in 1946 (*Croaking Movement*, *Earthworms* and *Eyes in the Heat*, part of a series he referred to as 'Sounds in the Grass') indicate some transgressions of this historically specific bureaucratisation of the senses. In addition, rhythm and movement can be experienced in several different senses, as witnessed, for example, in Pollock's titles for paintings such as *Autumn Rhythm (Number 30)* (1950), along with Harold Rosenberg's sensitivity to motion and rhythm. While Greenberg's criticism provided, in a sense, the 'official' and approved protocols for viewing abstract art, eyesight alone above them all, there was no guarantee that viewers, critics and artists were willing – or able – to follow the protocols. Suppressing multisensory responses to artworks requires exceptional discipline, the kind that is needed for retrospective accounts of experience that erase such 'noise'.

Many recent studies of mind wandering have shown that the sequencing of thoughts about the past, present and future is jumbled or rhythmically interleaved during such episodes. While this is going on, there may be interspersed moments of attentiveness to the external world, which in turn can structure these thoughts. Rhythmic and erratic cues in the external world prompt or even help to sequence sensations and conceptual associations. Thus, in following the knots and grain of wood with our hands, as Bachelard describes it, one could also be forming reveries, our fingers tracing the qualities of the material, forming images of rivulets, veins, threads, dragons or labyrinths. This is not a steady, conscious perception of a tree bark, which requires a particular focus, but a relaxed and passive receptiveness, an expanded perception of arbitrary textures and forms.

Bergson is interested in how this 'detensed' mental state becomes a 'duration' where past, present and future seem continuous. He writes that 'it is possible to imagine many different rhythms which, slower or faster, measure the degree of tension and relaxation of different kinds of consciousness' (Bergson [1896] 2004, 275). The perception of formless matter can easily trigger states such as pareidolia, where we are able to see a face in the clouds, or a twisted body, or a locomotive in a stain on the wall.[5] This is also expanded perception, but can we have perceptions without referencing prior forms held in the memory? Perhaps art's 'imageless' abstraction, as Kirk Varnedoe puts it (2006, 151), is a way of testing this.

The details Robert Morris's *Threadwaste* (1968) has to offer can hardly be called 'images'.[6] Bergson's detensed perception or Ehrenzweig's de-differentiation would involve the viewer not settling on any particular aspect of the artwork, and this is an important way to understand how abstraction works: it picks out low-level, general tendencies in shape, form or shadow that suggest rhythms and intervals. Paradoxically, it does not 'turn away from' (as in the Latin *abstrahō*) the artwork, but towards it. A relatively formless set of sensations seems to expand beyond percept formation. As the artist Georges Braque wrote, 'the senses deform and the

---

[5] Phenomena such as pareidolia may be a remnant of the brain's evolutionary development and may occur because the brain is designed to look for patterns. Gestalt psychology also provides some reasoning which might explain the way in which we instinctively order a chaotic scene. Yet artists have challenged the assumption that these efficient evolutionary mechanisms have much of a say in viewing abstract artworks. Jackson Pollock erased any part of his abstract work that could appear as faces or objects. Abstract art gives us permission to experience the purposeless pleasure of a disorganised composition. According to Anton Ehrenzweig, with de-differentiation we must learn to lift ourselves above such basic behaviour as reducing everything to faces in clouds if we are to appreciate abstract art.

[6] Although the viewer of *Threadwaste* may be reminded of a Jackson Pollock painting, particularly *Convergence* (1952), which contains a similar combination of colours and patterns.

mind forms' (quoted in Gooding 2001, 35). In Schopenhauer's terms, the will-less experience of the qualities of things-in-themselves emerges by lifting the veil of cognition that separates us from them.

Robert Morris writes that many artists are interested less in manipulating matter, to leave in it a trace of their personality, than in 'the more direct revelation of matter itself' (Morris 1993, 44). This is achieved by opening up a space in the mind where matter and gravity are allowed to run their course, where the 'focus on matter and gravity as a means results in forms that were not projected in advance' (44). This is what Morris calls an 'automated process':

Random piling, loose stacking, hanging, give passing form to the material. Chance is accepted and indeterminacy is implied, as replacing will result in another configuration. Disengagement with preconceived enduring forms and orders for things is a positive assertion. It is part of the work's refusal to continue estheticizing the form by dealing with it as a prescribed end (Morris 1993, 46).

With Morris's *Threadwaste*, perception becomes expanded rather than contracted into categories and order because categories and order are

Figure 1.1 Robert Morris, *Untitled* (1968). New York, Museum of Modern Art (MoMA). Felt, asphalt, mirrors, wood, copper tubing, steel cable and lead, 668 × 510.5 cm, variable. Gift of Philip Johnson. 504.1984. Digital image © 2020. The Museum of Modern Art, New York/ Scala, Florence. © Robert Morris/ARS. Copyright Agency, 2020. (A black and white version of this figure will appear in some formats. For the colour version, please refer to the plate section.)

suspended in the very first perceptual sweep.[7] This is not a psychology of observation that manipulates the geometric boundaries that belong logically to objects, and which it is possible to embellish with semiotics; that kind of 'intellectual' differentiation is suspended. Instead, we have a relaxed or distributed 'stuffing' strewn on the floor, faintly intuited remnants of objects or masses that once were, or material for new objects and textiles that we cannot anticipate. The material is implacably tangible and real, but also abstract and patterned in its chaos, with countless dizzying coloured threads.

Whereas 'consciousness attains to certain parts [of things] and to certain aspects of those parts' (Bergson [1896] 2004, 30–31), detension and de-differentiation support the kind of configural viewing that experienced viewers cultivate while looking at art (Cela-Conde et al. 2011, 45). For Bergson, detension is not simply a shift in perception but an ontological shift: rather than breaking off a bit of reality that is purposeful for one's needs, reality appears whole and one is part of it. It is this detension where the 'for itself' and the 'for me' seem momentarily to be indistinguishable. Even in rational reflection there is space for a question to arise: what if this fleeting wholeness, this objectless oceanic feeling with its fluctuations and vibrations, is less about the mind perceiving or differentiating matter and more about matter being itself? In *Creative Evolution* (1911), Bergson writes that the genesis of matter is not simply a mind/matter dualism because 'the mind part' consists of the organised intellect as well as the flowing and creative aspect of intuition.[8] For Bergson, intuition's elementary attachment to sensation makes it closer to matter than to intellectual form. For the philosopher Quentin Meillassoux, a fundamental problem in philosophy is the anteriority of matter to human cognition and the continual non-relatedness of the rational to this reality. For him, Bergson's detension can be understood as 'the operation through which Bergson "decontracts" the qualitative product of memory, so as to decant material perception from its mnemonic and subjective envelope – and this to rediscover matter such as it is in itself, rather than for us' (Meillassoux 2007, 80). But Meillassoux finds Bergson's understanding of pure perception, in which matter in itself can be experienced, as aporetic:

[M]emory-contraction seems to abolish the principal result of the theory of pure perception, namely that of the *cognisability* of the in-itself. For matter *appears to us* as that which has not been made the object of the work of contraction ... we cannot

---

[7] Figure 1.1 is a related work by Morris which closely resembles *Threadwaste*.
[8] Further unfolding this idea, Bergson writes: 'The universe endures. The more we study the nature of time, the more we shall comprehend that duration means invention, the creation of forms, the continual elaboration of the absolutely new' (Bergson [1911] 1969, 10).

see any convincing way to take the reverse path, so as to rediscover matter in itself not yet affected by our subjective duration (Meillassoux 2007, 82–83; my italics).

However, it is in our subjective duration that we become one with matter: we have a pre-reflective awareness of the mattered aspect of our subjective duration which seems entirely at one with the textures and rhythms we feel unfolding inside us and outside us. This does not mean that art is simply a process of private introspection (ego) about matter 'outside' (non-ego), but rather that it is only a traditional cognitive technique that divides art into internal and external domains in this manner, and this kind of knowing is not the only way for art to emerge.

The presupposition that cognisability is the only way that matter can appear as itself does not mean that cognisability allows matter to be itself. It is when the limits of cognisability are reached that anything outside it becomes un cognisable and, therefore, matter. In other words, matter-in-itself is where cognition ends and the terra incognita of the unprethinkable begins. In viewing art, this could be described as the optical unconscious or, in Ehrenzweig's terminology, de-differentiation. Modern psychology would describe this as the brain's metastable dynamics that are sensitised to the metastable dynamics of matter arranged by the artist in the artwork. This can be experienced as reverie, mind wandering or day-dreaming, relaxing cognitive restraint (cognisability). Bachelard suggests that the unconscious is closer to matter because it shares similar energies. For him, matter is the unconscious aspect of form: unconscious impulses 'flow unceasingly through our conscious life' (Bachelard [1948] 2002, 3) and engage us dynamically. For the philosopher Michel Serres, it is the id that is an undisciplined *material* force on the edge of chaos. Art presents matter-in-itself insofar as this presentation is received in mind-wandering episodes rhythmically entwined with the act of observation, released from the conscious supervision that would edit and parse these rhythms. It seems important that optical mastery, which we laud in art history as a methodical way of inspecting images and judging their aesthetic value, can be put aside in favour of a different frame of mind that delves into its own oneiric resources while rhythmically coupling with oneirically charged artworks.

For Bergson, the intellect is purposeful and analytical; we might say that it is good at decoding semiotics. Intuition for Bergson is rhythm, duration, sensation, without codes; we might say it is asemiotic. Bergson's intuition is not about distanced analytics but about entering into the thing intuited, as if intuition were one kind of matter in synchrony with another, and this involves being part of a multiplicity of rhythms that extend beyond the rules and distinctions applied by the intellect. The 'for us'/

'for itself' dualism can be elided in mind wandering and daydreaming, even if such episodes are introspective and autobiographical. In art, this 'for us'/'for itself' dualism can perhaps be grasped through the example of Cubism. When understood as 'for us', a Cubist painting seems to afford several possible viewpoints of an object, making us aware of our own place of viewing and our limited perspective and serving an epistemological function. The 'for itself' way of viewing suspends common sense and predetermined use value. The object seems to be indeterminate in *its* space, and the egocentric view switches to the allocentric. Both views are subject to ontological torsion, like the violin's neck and body that twist and turn under external pressure, and in our empathy and muscle memory the image may cause us to feel a certain contrapposto. It is as if the artwork shimmers between the 'for us' and the 'for itself', a duration in which art, literature and the imagination flourish.

Meillassoux writes: '[I]f matter is what Bergson says it is, then death – the return to the material state – would not at all be identified with nothing, but rather with madness – and even an infinite madness. For becoming-material would be the effacement of the selection of images' (Meillassoux 2007, 104). The dualism here is mindful self-reflection versus madness, but what kind of madness – mild schizophrenia or eccentric imagination? From the perspective of logical positivism, examples of creativity and art-making that draw upon mind wandering, daydreaming and imagining could also be called a kind of madness. This mind wandering is not 'a stupidity closed in on itself' (105), as Meillassoux suggests, but a rhythm between relaxing control and applying it, between creativity and evaluation, and this rhythm requires that one is able 'to maintain oneself in the Outside, but to hold oneself close, thus to some degree closed, and thus to discipline into writing a chaotic experience' (107). Niran Abbas refers to this as a 'turbulent, material poetics in which microscopic, unconscious perceptions constantly cascade into the realm of language, while, just as constantly, conscious perceptions become imperceptible' (Abbas 2005, 67). Mind wandering could be another way to describe this intertwining, where there are intuitions of oneiric rhythms in the coming and going of images and their textures that we have experienced in dreams. Yet mind wandering allows these dream rhythms to be ferried across into artistic practice and into the matter of artworks.

In *Matter and Memory* (1896), Bergson speculates that 'it is possible to imagine many different rhythms which, slower or faster, measure the degree of tension or relaxation of different kinds of consciousness' (quoted in Antliff 1999, 188). As Antliff explains and extends to the visual realm:

For this reason Bergson compares our consciousness to a 'melody'; similarly sensations such as colour are the 'qualitative' manifestations of the absorption of vibratory matter into the rhythmic tension of our duration. As the penultimate symbol of pure change, colour was a perceptual intermediary between the quantitative and qualitative, vibratory matter and the rhythmic pulse of our own consciousness (Antliff 1999, 188).

It is important that we remember how influential Bergson was for French artists in the twentieth century, not least because his ideas about matter folding and unfolding rhythmically in space and time were cooperative with memory and aesthetic experience.[9] These rhythmic pulses may be nonlinear, in the sense that a coherent narrative is not required or forced. This would mean that an openness to colour 'for itself' is prolonged to provide an expansive, slowed-down experience of time, the duration of colour, such as we see in the 'soft explosion' of a Rothko painting or in Dubuffet's *Matériologies*, which is enabled by avoiding measurements in tone. It seems important that Bergson's intuition is not purely idealism or realism, mind or matter, but an underlying continuity; it is not to be neatly segmented into past, present and future. Bergson attempts to characterise this continuity as vibration:

If you abolish my consciousness … matter resolves itself into numberless vibrations, all linked together in uninterrupted continuity, all bound up with each other, and traveling in every direction like shivers. In short, try first to connect together the discontinuous objects of daily experience; then, resolve the motionless continuity of these qualities into vibrations, which are moving in place; finally, attach yourself to these movements, by freeing yourself from the divisible space that underlies them in order to consider only their mobility – this undivided act that your consciousness grasps in the movement that you yourself execute. You will obtain a vision of matter that is perhaps fatiguing for your imagination, but pure and stripped of what the requirements of life make you add to it in external perception (Bergson [1896] 2004, 208).

What Bergson is suggesting here is both abstract thought, 'pure and stripped of what the requirements of life make you add to it', and relatively simple registration of the rhythms of matter, a visual experience the equivalent of which is feeling a pulse. It is important for Bergson that experiencing the vibrations of things, colour, sound, light is also to intuit that one's experience of these vibrations affect, and are affected by, the vibrations. To this intuition of a non-dualistic continuity, for which mind and matter are shown to be inadequate descriptions, is added the dimension of vibratory behaviour, pulsation, spasm. For Bergson, all life, objects, space-time and matter are complications of vibration. The

---

[9] For a good overview of Bergsonian influence on *Informel*, see Wilson (2013).

vibratory and rhythmic understanding of matter as 'delayed' or opened up by artworks is complemented by the notion of movement-matter. It seems important that not only are vibrations (or rhythms) difficult to fix or locate, they can also interpenetrate and propagate across the domains of art, body and brain. The very least that a basic understanding of nonlinear dynamics can do is to allow us to look at this complexity configurally rather than piecemeal.

Vibration also suggests haptic, felt, experiential qualities that are delayed or converted through the flesh: light literally enters the eye, and photoreceptors in the visual cortex turn these wavelengths of light energy into electrical signals and aggregates of sensation. Deleuze attempts to give an account of multiple kinds of matter-movement in his work on Francis Bacon's paintings, which he feels delay and help to complicate vibratory complexes of colours, sensations, and tactile and material variations. In allowing himself to be swept up in the sensations arising from his experiences, and yet being able to somehow relive their rhythms in his writing, he approaches what Meillassoux describes as maintaining 'oneself in the Outside, but to hold oneself close' (Meillassoux 2007, 107), so as to be able to write about a chaotic experience.

1.3   Recent Mind and Matter Complications

The scientist and theorist Katherine Hayles suggests that, along with particle physics and the idea of the subconscious, chaos theory is another post-Newtonian paradigm shift in how we understand natural and psychological phenomena and that it destabilises the very epistemological ground upon which positivist knowledge systems are built in both the sciences and the humanities. So-called new materialism seems to have gone to the root of this problem:

[W]hile scientific theories cannot simply be imported into philosophy, the tropes and rhythms they suggest can transform theoretical discourses. In fact, it is evident from new materialist writing that forces, energies, and intensities (rather than substances) and complex, even random, processes (rather than simple, predictable states) have become the new currency. Given the influence of classical science on the foundations of modern political thought, it is germane for new materialists to ask how these new conceptions of matter might reconfigure our models of society and the political (Coole and Frost 2010, 13).

Meillassoux argues that the material universe is characterised by chaos and contingency, and has no overarching reason to be as it is. For him, mathematics allows us to understand this 'mind-independent' reality, which also requires 'absolutizing mathematical discourse' (Meillassoux 2007, 80). In keeping with the mathematical interests of his mentor, the philosopher Alain Badiou, Meillassoux claims that mathematics is what describes the primary qualities of things, as opposed to secondary qualities which are manifested in sensory perceptions. On this view, mathematics might organise or generate itself as part of the thing described.[10] But while many would subscribe to the notion that chaos and unreason dominate the universe, it does not follow that recognising this stochastic universe 'for itself' can only be achieved through mathematics by which the 'for me' is somehow magically removed. There may be a 'non-cognitive route' to non-correlation in philosophy. Shaviro acknowledges such a route as

pre-cognitive (involving 'feeling' rather than articulated judgments) and involuntary (not directed towards an object with which it would be correlated) .... This new image of thought would maintain that aesthesis, or precognitive feeling, precedes noesis, or cognitive apprehension. ... Nonconscious experience is not an oxymoron; it's simply that more things are felt than are known (Shaviro 2013, 56).

I believe that the complications of mind wandering provide a good example of what this non-cognitive state might look like.

This is somewhat comparable with the 'cognitive unconscious' proposed by Hayles, which consists of somatic knowledge, sensorimotor competencies that pass underneath consciousness. Both Shaviro's and Hayles's approaches help us to go beyond the narrow view that all cognition is intentional, logical, computational or detached from the world. Heidegger's *Dasein* and the ready-at-hand established this principle of being absorbed in an activity where the distinction between the 'for us' and the 'for itself' dissolves. The 'cognitive unconscious' allows us to be competent or useful in our daily tasks such as driving a car or cleaning the house. Although mind wandering is often understood as withdrawal from the world for introspection, it can also be seen as an engagement with the earth, materials and their shape-shifting patterns, which we experience through our attunement to abstraction. We privilege rational analysis of

[10] Johnston (2013) accuses Badouian philosophy of being too closely connected to the mathematician Georg Cantor, whose achievements form only a subdivision of formal science, which Johnson describes as being

based on false, out-of-date impressions of how nature and matter are conceived of in biology and related fields (in this regard, Badiou-the-philosopher has not lived up to his philosophical duty thoroughly to think his contemporary scientific circumstances) ... [furthermore] Badiou's reasons for insisting that, of the sciences, only pure mathematics is suitable for ontological purposes due to its supreme abstractness are reflective of Platonic-idealist tendencies to devalue real material existence in positing a purified being-in-and-of-itself. Separating out a clean ontology of being *an sich* from the dirty, messy onticness of concrete entities, regardless of how this separation is executed, always brings one into the proximity of antimaterialist elements (Johnston 2013, 107).

objects in the world as a superior form of engagement with the world, but rational thought is also an engagement with memory, learned processes and schemata, procedures which we superimpose over a metastable visual field. It is worth thinking about what such a being-in-the-world would be like if we were to delay these procedures in order to see unstructured aspects of the world – the gold-flecked patterns on the surface of a painting or embedded in the paper of an illuminated manuscript – regardless of meaning. This brief moment, an infinity in a grain of sand, is not just an irrelevant blip. It is a being-in-the-world where the nonlinear dynamics of sensations, memories or feelings are cued or accented by the starts and stops and complex play of the abstract patterns we are looking at. It is art that helps to reinvent this way of being with matter.

Heidegger speaks of the enigma and mystery of materials in excess of cognitive mastery, and this idea continues in the work of later philosophers such as Graham Harman. He writes, for example, that 'no matter how exhaustive our knowledge of the moon may be, that knowledge *is not itself the moon*' (Harman 2015, 148). Apart from the fact that only an obsessive cognitivist approach would insist on counting every energy particle of the moon in inconceivable levels of granularity in a failed attempt to 'know' it, Heidegger's letting things be (*Gelassenheit*) suggests that the relinquishment of the cognitivist stance, the will to know, *at the same time* allows the moon to come into being. 'Letting be' is letting things unfold into their being-ness, which becomes part of our being-ness in letting be. This is perhaps better described as a non-cognitive multisensory attunement with the textures, tones and patterns of an object without having in mind a strong subject/object distinction or even an explicit thought about the object as an object to be quantified. Abstract art encourages this temporary forgetting of objecthood, where we participate in the metastability of matter through the metastability of our sensations.

On this view, art restores to us the space and time for a vital connection to nonlinear, dynamic and itinerant *Dasein* or being-in-the-world. It may also provide us with an attunement to the rhythms of uncountable matter – trees, clouds or earth freed from the boundaries we draw around them to distinguish objects from the subjects observing them. The moods and sensations are not only attuned to these rhythms but also extend them across brain, body and world. Heidegger uses the term *Grundstimmung* (basic mood, tone and attunement). We can understand this as an example of the attunement of brain, body and world, which I would suggest is the fundamental philosophical kernel in Ehrenzweig's more psychological understanding, where 'outward perception and inner phantasy become indistinguishable' (Ehrenzweig [1967] 1993, 272).

This indistinguishability is the moment when nature, the artwork and self-reflection disappear in reverie.

Another way in which the stark contrast of matter and mind, subject and object, appears to become a continuum of sorts is offered by the concept of a fold. A fold creates a distinction within the same entity, as opposed to a cut, which divides into a dualism (or what is assumed to be a dualism) that needs to be joined or sutured. On this view, one way to free oneself from rumination about the 'for itself' and the 'for me' would be to accept an underlying monism where folds and differentiations appear only through certain philosophical problems and common-sense 'extractive' perceptions. The virtue of a fold is that it can become unfolded. Thinking in terms of folds and unfolding requires cultivating a psychology of reverie. When art is seen as a fold in being, it becomes possible to imagine that it can be 'unfolded' to include the viewer's mind. This requires a fundamental openness (unfolding) to the artwork's dynamics, which are not folded into 'me' and 'it' distinctions. Mind wandering temporarily smooths these folds, these distinctions, into the plane of being-in-the-world.

Shaviro suggests that panpsychism might offer a way through the problems raised by Meillassoux:

[I]f we accept that thought (or feeling or experience) need not be conscious, then we might well be led to abandon the demarcation between mind and matter altogether. ... I propose that [panpsychism] gives us a good way to avoid the problematic baggage both of consciousness and of phenomenological intentionality (Shaviro 2013, 44).[11]

For Robert Jackson, the co-relationist divide can be bridged by high modernist art:

[P]resentness within an artwork operates when the work deliberately withdraws from an explicit method of relational execution and never attempts to seek out the beholder's presence, whilst at the same time being paradoxically constructed by the beholder's anthropomorphic finitude. For Fried this manifests itself in beholders being entirely absorbed in their actions, feelings and thoughts, completely unaware of their situation vis-a-vis the artwork (Jackson 2013, 349).

Matter painting is more successful at avoiding some of the cognitive bias and intellectual baggage of high modernism. For Jackson, what is crucial is

the ability for the aesthetic effect to occur in a non-anthropocentric manner, without the acknowledgment of a human beholder; i.e. the primordial logics of projection, absorption and presentness occur between neighbour cat and catnip,

---

[11] Meanwhile, Protevi speaks of affective cognition as a way to suggest material foundations for cognitive thought instead of panpsychism (Protevi 2009, 74).

sodium and chloride, lemon juice and salmon flesh, wheat and flame, as well as the work of art and beholder (Jackson 2013, 357).

Jackson wonders whether it would be possible that

a twenty-first century aesthetics will not be about a return to objects in the literal sense, but about a realist explosion of isomorphisms: any mindless, inanimate, cephalic or a-cephalic beholder in its withdrawn inwardness has the potential to be, in a realist manner, absorbed through its own projected presentness of withdrawal (Jackson 2013, 358).

Again, the problem with many kinds of high modernist artworks is that they address and service cognition and intellectual constructs. They are meant to reflect these back to us and do so with remarkable clarity, particularly geometrical abstraction, which mobilises measurement, scale and other intellectual categories. The material processes that Jackson proposes might be unleashed in a non-anthropocentric isomorphism of the 'for us' and 'for itself', not in the geometrical abstraction of high modernism but in Morris's *Threadwaste*, Smithson's artworks and *Informel*. It is the intellectual formalism of geometrical abstraction, balance and order that leads back to Kantian detachment and the cognitive procedures that are further removed from matter. This suggests that formalism is normally understood as a rational appraisal process that requires a Euclidean common-sense grasp of geometrical forms. Yet in Morris's art and in matter painting we see that these forms are continuously being undone, and formalism here is about loose aggregates and fine-grained textures rather than hard edges.

Other ways of explaining the gap between mind and matter, the 'for itself' and the 'for me', externalism and internalism, are suggested by Adrian Johnston, who notes that his project centres on

a gap between, on the one hand, a detotalized, disunified plethora of material substances riddled with contingencies and conflicts and, on the other hand, the bottom-up surfacing out of these substances of the recursive, self-relating structural dynamics of cognitive, affective, and motivational subjectivity – a subjectivity fully within but nonetheless free at certain levels from material nature. I wager that the genesis of this very gap can and should be explained in precise detail through a transcendental materialism drawing extensively upon Freudian-Lacanian psychoanalysis and the life sciences in addition to philosophy past and present (especially German idealism, Marxism, and Anglo-American analytic philosophy) (Johnston 2013, 208).

This gap is supported by metastability in the brain, open to the equipossibilities of perceiving the details of matter painting and abstraction.

The issue of mind and matter in philosophy is often stated in terms of the controversy between 'substance dualists' and the non-dualism of materialists (further subdivided by many different positions and arguments). A good summary can be found in Churchland's *Matter and Consciousness*, where he correctly states, '[T]here is no doubt at all that physical matter exists (and plays a substantial role in our internal cognitive activities), while spiritual matter remains a primitive, tenuous, and explanatorily feeble hypothesis' (Churchland 2013, 29). Churchland places great emphasis on neural networks playing a significant role in how consciousness is physically produced. His approach has been consistently associated with 'eliminative materialism'. He argues that a wide variety of mental constructs can be explained through scientific investigation into matter and material processes, even if, at present, it is difficult to do so. Although I agree with Churchland that all mental states are physical states, I am wary of the term 'states' and the kind of neuroscience he deploys, particularly forms of reductionism, where bits of brain are matched to a 'state' to provide evidence that mental states are physical states. This kind of reductionism does not capture the bigger picture of the brain's dynamic complexity and chaos, which is physical and in flux.[12] György Buzsáki's *Rhythms of the Brain* (2006) is a powerful antidote to the traditional reductionist paradigm.[13] For him, a so-called mental state is better understood as a snapshot of continuous and interwoven brain-body-world interactions with multiple-level activities taking place in each domain. It seems that the best way to give an account of multiple-level complexity is through nonlinear dynamics and by linking levels of description, or 'intertheortic' relations as Churchland puts it.[14]

---

[12] Complementing this is Merlin Donald's observation concerning the limits of reductionist neuroscience, that it 'would not be unlike a particle physicist's trying to track every electron in, say, a roomful of people in a cocktail party. Why would one want to do this? It would explain nothing about cocktail parties or people. Nor would such an analysis explain a work of art' (Donald 2006, 13). Kandel (2016) attempts an analogy between scientific reductionism and a simplistic potted history of abstract art (with glaring omissions such as Malevich), based largely on the notion that abstract art eliminates figural representation so that various forms of pattern completion or associative recall are activated. But this is using 'reductionism' in a overly simplistic manner for art. Only some forms of abstraction can be seen as reducing details, such as with Piet Mondrian, but even here, there is much more to his work than perceptual processing. In other cases, the elimination of the figure does not reduce detail; rather, the opposite happens: abstraction allows thoughts and sensations to become more complex and interwoven. In a painting by Gorky, Gottlieb, Krasner, Tobey, Pollock, Frankenthaler, Miró or countless other abstract painters, abstraction is more properly understood in terms of dynamic complexity, chaos, phase transitions and relative stabilities.

[13] I do not believe that this puts me in the substance dualist camp. Neural populations, brain oscillatory frequencies and networks in continual activity with a continuously eventful environment are all matter, but dynamic and complex matter in movement in relation to non-brain material.

[14] There may be similarities with the functionalist argument explained by Churchland, whereby mentality 'is not the matter of which the creature is made, but the causal

Churchland's eliminative materialism proposes that consciousness will become transparent through the discoveries of theoretical and empirical science and has built within it a positive stance towards futurity. But if for him 'conscious intelligence is the activity of suitably organised matter' (Churchland 2013, 261), we might also ponder different kinds of organisation, such as the more metastable kind producing mind-wandering activity, which, although not entirely conscious, is essential for our well-being and creative endeavours. Churchland writes that 'we are creatures of matter. And we should learn to live with that fact' (Churchland 2013, 35). I believe abstract art and 'matter painting' provide this proposition with insights that help us to realise this and provide experiences of it. Both Bergson and Bachelard suggest that the mind is closer to matter when it relaxes reason and becomes lost in daydreams. Art reminds us of how matter outside the brain can entrain or surprise us, leading to learning. Matter creates newness in the world in excess of what we 'know', and in this sense it is a kind of futurity. Perhaps this aspect is addressed by the claims of eliminative materialism, but with one difference: futurity is one of the most important things the artist works with in a fine-grained way, with matter moving unpredictably between bodies. This is so important, in fact, that the artist is willing to risk not knowing in order to help produce it, skilfully avoiding the kind of consciousness that forestalls experimentation, the unprethinkable and the freedom of the imagination.

It may be an idealist claim that humans continually seem to outperform their evolutionary neurological and cultural programming, yet it is this outperforming that is always one step ahead of reductionist explanations that attempt to grasp it. The stronger claim is that this continual outperforming, which appears to be in excess of the material substrate that produces it, is part of the very nature of matter, its virtual edge. While eliminative materialist approaches build rational theoretical structures

[14] structure of the internal activities which the matter sustains' (Churchland 2013, 35). This has meant a lot of research into whether the functional organisation of certain groups of neurons can be reproduced in computers to arrive at the same mental state (or consciousness), and some of these research paradigms are used to determine metastability in the brain. However, it may be that certain internal activities occur in particular strengths of connection or combination because of the matter itself and its particular history. Mixed or nested brain frequencies, for example, may come together in particular complex routines in engaging with an artwork and may add novel features to the so-called mental state identified. In other words, we can point to coarse-grained mental states and be generally talking about the same things across different cases but there is always a fine-grained level of idiosyncratic 'noise' which may be crucial in subsequent developments. Eliminative materialism may eventually be able to ascertain the general importance of this aspect but it can hardly be expected to give an account of every singularity or every kind of action potential or intention not fully passing into consciousness or becoming a properly identifiable 'event'.

around the rift of futurity, art seems to dive straight into its depths to find something lasting.

1.4   The Matter of the Brain

Bonta and Protevi write:

[J]ust as Kant's *Critiques* were in a sense the epistemology, metaphysics, ethics and aesthetics for a world of Euclidean space, Aristotelian time, and Newtonian physics, Deleuze provides the philosophical concepts that make sense of our world of fragmented space ... twisted time and the nonlinear (Bonta and Protevi 2004, iv–viii).

John Marks claims that Deleuze and Guattari build what has been referred to as new materialist philosophy, which attempts to grasp the self-organising properties of 'matter-energy'

Deleuze finds, in thinkers like Spinoza and Simondon, a mode of thinking that challenges the dominant Western notion of matter as inert, passive 'stuff' that requires an external form in order to exist in a concrete and recognizable way in the world. Spinoza, for example, raises the possibility that matter might not require the external imposition of form; it might be that the capacity to take on a form is immanent to matter. That is to say, matter might be capable of morphogenesis. ... Following Bergson, Deleuze and Guattari start from the assumption that matter is best thought of as energy, rather than as mass, and they emphasize what they see as the creative capacities of matter (Marks 2006, 4).

Deleuze and Guattari's understanding of matter 'encourages the view that extensive material structures and objects that we see around us are simply "coagulations or decelerations" of intensive flows of matter-energy that comprise the material world' (Marks 2006, 4). It seems essential, then, that the brain is also seen as a complication of these intensive flows of matter-energy.

Marks writes that this Deleuzoguattarian understanding of matter is in step with the science of complexity and emergence, emphasising the active, self-organising properties of 'matter-energy'. Catherine Malabou (2008) notes that Deleuze is one of the few philosophers of his time to have taken an interest in neuroscientific research. She suggests that the way in which Deleuze brings together neuroscience and philosophy is by focusing on the brain as an acentered system. Both Deleuze and Malabou concentrate on the breaks and gaps of the synapses to show how neural plasticity proceeds by leaps that are not planned or preconceived. Malabou writes of 'synaptic gaps' that play a key role in cerebral organisation:

Nervous information must cross voids, and something aleatory thus introduces itself between the emission and the reception of a message, constituting the field of action of plasticity ... the discovery of a probabilistic or semi-fortuitous cerebral space, 'an uncertain system' according to Deleuze, implies the idea of a multiple, fragmentary organization, an ensemble of micro-powers more than the form of a central committee. In consequence, our lived relation with the brain becomes more and more fragile, less and less 'Euclidean' (Malabou 2008, 36).

Buzsáki also explains that loops of activity in the brain

are not closed by brain wiring, but there is a 'gap' between the neuronal connections controlling the outputs and inputs that transmit information from the sensors. The gap may be closed by actions exerted by the brain on the body and the environment, a process that 'calibrates' neuronal circuits to the metric of the physical world and allows the brain to learn to sense (Buzsáki 2006, 36).

It is to Deleuze's credit that the 'brain, body, environment' approach is discussed in terms of cinematography and art rather than shopping or doing the laundry. A number of commentators on Deleuze make this clear. Malabou writes that, for Deleuze, Alan Resnais is a 'filmmaker of the brain' who produces 'landscapes' that are 'mental states' (Malabou 2008, 17).[15] John Mullarkey notes that Deleuze's cinema books are 'inspired by the microbiology of the brain' and that 'Deleuze sees cinema as operating at the molar level to shock the brain into forming new synapses, connections, and pathways' (Mullarkey 1999, 63). According to Andrew Murphie, Deleuze suggests that montage in film, a technique of closing gaps between images, creates 'a set of harmonics acting on the cortex which gives rise to thought' (Murphie 2010, 62). He also suggests that, for Deleuze and Guattari, the structure of the brain 'is not so much "syntactic" as topological, a probabilistic series of chaoses in interaction, which are folded by ongoing connections between "extreme elements"' (62).

These approaches are interesting because they help us to form a picture of the brain not as an isolated entity but as something that is actively engaged with, and part of, the world of matter in all its variety, transformations, signals and rhythms, some of which seem to be 'articulate' and to talk back, as we see with film and art. Art can remind us that the brain is a fold (or many folds) in matter, even as it is able to think about and imagine these folds. Its operations are filtered from the outside world insofar as its own self-organisation is not directly subject to every stochastic perturbation in the world, although scientists maintain that 'complex bodily rhythms are ubiquitous in living organisms. These rhythms arise from stochastic, non-linear and biological mechanisms interacting with a fluctuating environment' (Sarbadhikari and Chakrabarty 2001, 448).

---

[15] Pisters (2012) develops the Deleuzian premise that the cinema screen acts as a kind of brain.

Deleuze and Guattari speak of the brain as a junction. This suggests that the brain can pass on some part of the signals it receives, or transform them, switching direction and speed, adding to and elaborating signals, or preventing them from being passed on by choosing one set of signals over another. Neurons have multiple connections with other neurons, depending on the task they perform, and they work in orchestrations of oscillatory rhythms.

Deleuze and Guattari are remarkable for anticipating in the 1980s some of the important conclusions reached by neuroscientists today. But why turn to the philosophy of Deleuze and Guattari if modern neuroscience provides more details and precision about the brain? Because it is through their work that the (non-reductionist) neurodynamic brain can be understood within the context of the body and the world, which are also dynamic, affecting and being affected by art and film. Recent studies reveal that the brain not only functions at the level of electrical signals between neurons, or as regions or networks with particular functions, but is also an extremely complex set of oscillatory frequencies premised on metastable dynamics, often described as 'chaos in the brain'. It is interesting that there are no studies of the metastable dynamics of the brain when it is involved in viewing abstract art. For Deleuze and Guattari, art, philosophy and science have their own particular ways of working with chaos, but the brain, which processes these different systems of thought, also 'composes' chaos:

> It is up to science to make evident the chaos into which the brain itself, as subject of knowledge, plunges. The brain does not cease to constitute limits that determine functions of variables in particularly extended areas; relations between these variables (connections) manifest all the more an uncertain and hazardous characteristic, not only in electrical synapses, which show a statistical chaos, but in chemical synapses, which refer to a deterministic chaos (Deleuze and Guattari 1994, 213).

The brain functions chaotically in its metastability and openness to the chaos that is part of the world outside, but it is also highly organised, as are the structures found in the material world. The brain is able to adapt to this range between chaos and order, this composition of chaos that artworks present. Within this context, the artwork is not simply a fixed or finished product but something that helps to stabilise or destabilise an event distributed across domains. The artwork is a nonlinear composition of the movement-matter of vibrations and rhythms, switches and delays, and helps to produce nonlinear sensations and movements of the body and the neural assemblages of the brain, which in turn may select certain features of what is seen. This nonlinear feedback mechanism is shared by the brain-body-art assemblage. This assemblage is not bound to produce harmonies, narratives or meaningful passages reflecting traditionally revered structures of order; thought, action and art can work together to produce a metastability, creating new connections.

Neuroscientist Kelly Clancy suggests that Deleuze and Guattari's idea is that the brain is like an umbrella protecting us from chaos: 'Order and disorder enjoy a symbiotic relationship, and a neuron's firing may wander chaotically until a memory or perception propels it into an attractor. Sensory input would then serve to "stabilize" chaos' (Clancy 2014). The suggestion here is that the rhythms of external events can act as less-than-conscious attractors for a systemwide change or 'phase transition',[16] such as we see in mind-wandering behaviour or meditation. This is in step with Deleuze and Guattari's understanding of the brain and its relations with the world. Deleuze and Guattari are sensitive to a nonlinear, dynamic and configural understanding of the relations within the brain – and between brain, body and world – variable and shifting relations that are both continuous and discontinuous.

1.5   The Matter of Sensation

> Sensation refers to the initial processes of detecting and encoding environmental energy . . . potential energy signals from the environment emit light, pressure, heat, chemicals, and so on, and our sense organs . . . receive this energy and transform it into bioelectric neural code sent to the brain. The first step in sensing the world is performed by specialised neural units, or receptor cells . . . that react to specific kinds of energy
>
> (Schiffman 2001, 2).

This is another example of how fine-grained scientific details contradict the substance-dualist notion that mind and matter never comingle. We know, for example, that cells in the tongue react to chemical molecules, and olfactory sensations arise from physical contact with the particles of the thing we smell. These processes form a continuum between the internal and external that is the material substrate of sensation.

Where does sensation begin and end? This is particularly difficult to isolate once we understand that sensation is a dynamic process, continuously refreshed by input and stimuli from the environment. On this view,

---

[16] In order to understand criticality, one needs to become familiar with phase transitions. A 'phase transition' is the process by which one state or 'phase' of matter transforms into another: water, through condensation or freezing, and by analogy, the brain in a state of alertness or drifting off into sleep. The moment of transition from one phase to the other is called the critical point. It is useful to think of 'criticality' as a duration, or even a quality, in its own right rather than on its way to becoming something more important. This may also be a useful way to understand the duration of mind wandering as being, which some art helps to open up for us.

isolating sensation as a phenomenon immanent only to the body makes no sense. It is worth keeping in mind the following:

1. Physical energy (variable intensities of light) enters the eye, is reflected on the retina as a luminance pattern and transduced into electrical impulses which are distributed by neural pathways to brain areas, allowing us to perceive such things as motion, luminosity and outline.
2. These basic patterns can be filled out and elaborated by motor and haptic routines to help produce sensations of texture, sharpness, tone, bitterness, sweetness, warmth, softness, rhythm.
3. This can be elaborated to an experience of things and objects by further perceptual processes, and semantically with cognitive processes triggering stored inferences and percepts of objects.

Many abstract artists hope to extend (1) and (2) in their art, often attempting to undermine or delay (3). These artists, and perhaps philosophers of sensation such as Deleuze, would understand that artworks are able to 'return' us to (1) and (2) even if (3) intervenes. In fact, in *The Logic of Sensation*, Deleuze suggests that sensation can be understood as a material spreading effect that stretches across the artificial distinction between the internal and the external, even though it changes its material substrate as it does so.

Physical energy patterns should not be understood as raw material, worked on and refined by later brain processes that are a step removed from 'the physical', but rather as increasingly complex extensions (or abstractions) of the physical. Viewed in this way, sensation could be seen as a liminal zone between disorder (raw physical matter) and order (perception). In this understanding, sensation connects disorder and order. And Deleuze certainly believes that the sensations encouraged by Francis Bacon's art sustain this connection. Bacon's half-abstract, half-figurative style helps to sustain the metastability of sensations arising from the raw impasto and the perception of a face.

Often in Deleuze's work, sensations are implicitly associated with the vocabulary of emotion or, more precisely, affect. But by replacing the logic of direct perception, which assumes that our stored percepts are mirrors of things in the world as they are, 'direct sensation' preserves the notion that we are coupled in a biologically relevant structural way to the rhythms of matter (whether in the form of light, vibration, texture or tendencies) even if at some stage perceptions take us away from them.[17]

---

[17] Direct perception has often been compared with 'naïve realism', which holds that we are 'more or less in contact with the world and that what we perceive by and large is the world as it really is' (Mausfeld 2010, 161). Even though Mausfeld convincingly rejects the notion that perceptions mirror the world 'as it is', he concedes that perception must 'at least not contradict biologically relevant aspects of the external world' (161). This non-contradiction could be manifested as rhythm.

Sensations, richly intertwined and transmodal, which art helps to complexify, are not stochastic but 'a composed chaos', a metastability that, when seen as a nonlinear dynamic process over time, goes back and forth between (1) and (3), with (3) being increasingly ordered and (1) being chaotic. Sensations take us *away from* or *towards* the zero point of the purely stochastic, to which we may connect directly, but always at the risk of no return the longer we remain with it. Art can be seen as something that draws us towards (1) in some kinds of matter painting and abstract expressionism, or closer to (3) in some conceptual art.

It is the brain that plays a key role in the modulation of sensations, in conjunction with the body and the world with which it comes into contact. Deleuze's work on Francis Bacon takes as its subject sensation and how it is dynamically sustained across brain, body and art. In this sense, Deleuze's approach has psychological, phenomenological and philosophical dimensions. A good summary of how Deleuze might have been influenced by neurologist Erwin Strauss's understanding of sensation is given by Daniel W. Smith in the preface to Deleuze's *The Logic of Sensation*. Smith explains that, for Strauss and, by extension, Deleuze (and Bergson), perception is different from sensation precisely because it has a tendency to order and constrain sensation, to support orientation and boundaries of self and other. Yet Smith writes that 'the prerational world of sensation is not *prior* to the world of perception or representation but ... is *coextensive* with it' (in Deleuze 2003, xiv).

For Deleuze, sensations in their many aggregates of rhythm are the direct, non-rational and intensive connections we have with matter-movement – a qualitative *immanence* of rhythmic vibrations across brain, body and world that may also provide a more intuitive 'felt' connection with the world rather than extracting immediate opportunities for survival. Much of *The Logic of Sensation* is given to exploring how colour in painting creates puzzles and intensive tactile, optical and rhythmic sensations that seem to unfold in nonlinear ways. If colour is also sculptural and can spark conceptual associations, as well as being a biological process with emotional effects, it stands to reason that the multiple effects of colour can be pursued through different disciplines such as neurobiology, phenomenology, psychology and even the physics of refractions and reflectances of colour scattered or concentrated by brushstrokes. In this way, Deleuze's approach to art suggests a nonlinear dynamics of sensation that is sustained across brain, body and world, and which is best understood through a multidisciplinary approach.

The idea that sensation involves brain, body and world across internal and external considerations is expressed by Deleuze when he writes:

Sensation has one face turned toward the subject (the nervous system, vital movement, 'instinct,' 'temperament' ...), and one face turned toward the object (the 'fact,' the place the event). Or rather, it has no faces at all, it is both things indissolubly ... at one and the same time I *become* in the sensation and something happens through the sensation, one through the other, one in the other ... the unity of the sensing and the sensed (Deleuze 2003, 31).

In *The Logic of Sensation*, Deleuze suggests that not only can sensations across the five sensory domains share rhythms, but the objects of such sensations in each domain (the music of the auditory sense, the painting of the visual sense, the surface of a rock for the sense touch) may share rhythms as well. There are also countless nested rhythms in the brain and body, which suggest that mental states depend on metastable dynamics rather than functionally stable brain areas or faculties. Deleuze is interested in examining the semi-chaotic reverberations of rhythm across brain, body and art, which I would associate with a description of Bernini's *Ecstasy of St. Teresa*.

The convulsions, spasms and paroxysms of Bacon's anatomies (Figure 1.2) are diagrams of 'heads whipped by wind, deformed by an aspiration – but also all the interior forces that climb through flesh. To make spasm visible. The entire body becomes plexus' (Deleuze 2003, xxix). Bacon's paintings of heads hold a special fascination for Deleuze:

[T]he extraordinary agitation of these heads is derived not from a movement that the series would supposedly reconstitute, but rather from the forces of pressure, dilation, contraction, flattening, and elongation that are exerted on the immobile head. They are like forces of the cosmos confronting an intergalactic traveller immobile in his capsule. It is as if invisible forces were striking the head at different angles (Deleuze 2003, 50).[18]

It is here that we see how Deleuze's approach takes on multiple levels of description from gravity, velocity, force, the different angles of relativity, the different material and rhythmic sensations, 'archives of the flesh' engaged in contortion, dilation, flattening and elongation of the paint, gesture – all centred on the head. These 'lines of force' and movement are not really representations that a detached viewer might descry, for they help to produce, with their various tactile qualities of resistance,

---

[18] Cupchik proposes that metaphorical concepts such as 'forces' and 'fields' should not be taken too literally lest they become reified: 'The focus should be placed on the dynamic effects of compositional contrasts and resulting tensions rather than on abstract "forces"' (Cupchik 2007, 16). 'Forces' remains a useful term for abstract painting and matter painting, used to highlight or generally appreciate a directional tendency towards becoming organised, or dissipating and becoming disorganised or disaggregated. In this sense, 'force' can be understood as leading to or from some relatively identifiable state. But Deleuze would insist that feelings and sensations also create and are subject to forces in their intensities or causal effects in the world, and in terms of particular expenditures of energy that the artwork is able to inspire.

Figure 1.2 Francis Bacon, *Study for Portrait, Number IV (from the Life Mask of William Blake)* (1956). Oil on canvas, 61.5 × 51 cm. Digital Image © Museum of Modern Art, New York/Scala, Florence. © The Estate of Francis Bacon. Copyright Agency, 2020. (A black and white version of this figure will appear in some formats. For the colour version, please refer to the plate section.)

sensations of some force in the viewer's haptic and kinaesthetic feeling. Bachelard's description of the muscular lyricism of Lautréamont could easily cooperate with Deleuze's description of Bacon. Bachelard uses phrases such as 'images of propulsion' and 'hierarchy of velocities' to suggest an interest in producing speeds, rhythms and sensations, and not merely referring to them.

This dynamic immediacy could more accurately be described as participation. According to Deleuze, the matter painting of *Informel* extends

the diagram to the entire painting, it takes the diagram for the analogical flux itself, rather than making the flux pass through the diagram. This time, it is as if the diagram were directed toward itself, rather than being used or treated. It no longer goes beyond itself in a code, but grounds itself in a scrambling (Deleuze 2003, 81).

Rhythm allows human matter to affect and be affected by non-human matter by way of micro or macro vibrations, synchronies or 'translations' – a 'vibrating together', as Bergson expressed it. The vast number of different kinds of matter in and through which these vibrations occur, and which are swept up into rhythmic patterns and sensations, requires us to diversify our vocabulary of vibration, where it would be more appropriate to speak of flickers and fluctuations for light and energy, wavelengths for colour, wave forms for sound and for water and its cognates, oscillations for the brain, sedimentary disturbances for the earth, *et cetera*. In Deleuze's understanding of art, sensations 'resonate' and are 'coupled together like "wrestlers" ... there are also resonances that are derived from the layers of superimposed sensations' (Deleuze 2003, 59, 61), and Deleuze examines Bacon's triptychs to show how this layering, wrestling struggle between sensations, resonates. Bacon's work is described in terms of consistency, coagulation, elasticity, flux, torsion, spasm, struggle, resonance, force, a diastolic-systolic opposition (which brings to mind the rhythms of the heart), and ascending and descending movements, with references to the ancient mythos of the clinamen, expenditures and discharges of energy, and 'the fall' – the latter being particularly relevant for the way Bacon was known to struggle with the Christian tradition, where the fall from grace is a major theme. The consequence for these kinds of combinatorial, mobile engagements with Bacon's art, particularly the triptychs, is an art historical approach based implicitly on nonlinear dynamics. Deleuze writes of

vibrating sensation – coupling sensation – opening or splitting, hollowing out sensation. These types are displayed almost in their pure state in sculpture, with its sensations of stone, marble or metal, which vibrate according to the order of strong or weak beats, projections and hollows (Deleuze 2003, 164).

This formulation taps into the rhythms of sensation as they form into percepts, and the authors attempt to capture the metastability that this creates, particularly because they avoid the suggestion that this occurs as a linear unfolding. They creatively weave together the multi-temporal strands and lines of force of the figures, and the application of colours and their spreading effects, through their own intricate writing that avoids 'telling a story' with a beginning, middle and end.

Bacon's compositions cue percept-accented sensations, where the coarser forms reference 'forces' and fluxes, ruptures and instabilities, while the fine-grained details disperse these distributions further. There are many different kinds of stimulation and vibration at the fine-grained level that feed into the coarse-grained view: edge detection is intricate and dense, interwoven with strands of colour; there are countless mental shape rotations, vertical and horizontal thrusts with some diagonal aberrations; and there are various tensions in figure-ground ambiguities. These complications do not function simply as tools to fix the truth of 'the image' but are dynamic mental events cued by but not identical to the composed chaos found in the painting. Thinking about the timescales and rhythms that prefigure, postdate and are nested within any event is a difficult exercise in nonlinear dynamics, which is made easier by viewing a Bacon painting. Here it is possible to project rhythm and topological patterns onto a static medium, as we would with simple mental rotation, and the painting's details stimulate such projections.

This approach, sensitised to nonlinear dynamics rather than the linear narrative of telling a story, is particularly relevant for painting because it is a nonlinear medium: the 'entry point' into the pictorial space or the order of 'reading' from point to point is not prescribed and the 'syntax', the order of features focused on, may be different every time a painting is viewed. The canvas is indeed 'infinite'. But with Bacon, whose figures, lines and points seem to twist and turn depending on where one begins one's visual journey, it becomes possible to visualise the emergence of a dynamically elastic three-dimensional modulation of sensations involving the artist's supple gestural and nervous movements, the material elasticity of the pigment, and the viewer's own understanding of how things are achieved through movement, pressure and rhythm. One can read across the canvas the artist's hand movements, twisting, quick or hesitant, erratic or tender, the brush loaded with a spectrum of colours that are blended in the gesture of applying paint, when the hairs of the brush come together as they leave the canvas.

In Deleuze and Guattari's *A Thousand Plateaus* (1988), sensation becomes much more philosophically grand, for its rhythmic components can not only be shared across sensory domains and organs, paintings and music, but can also be co-opted and exploited by social systems of regulation. What we see as natural sensations are conditioned by centuries of cultural and political organisation. If sensations are continuities across brain, body and world, one of the ways in which they could be understood to extend across so-called internal and external worlds, and transmodally, is through pulse, rhythm and pattern. As the various oscillations of the brain indicate, we are receptive to rhythm in unconscious, semi-conscious and fully conscious phases. Bachelard extends the concept of rhythmanalysis based on the Brazilian philosopher Pinheiro dos Santos's work on digestion, the unconscious, love, morality and mood.

'Rhythmanalysis' is also the title given to Lefebvre's work enlarging his analysis of the everyday. For Lefebvre, capitalism produces certain rhythms of production that, unbeknownst to us, influence our movements and choices in built environments and in the routines of our everyday lives. On this view, the everyday functions as an embodied and social attunement to the clockwork rhythms structured by family, school, consumerism and work. We automatically synchronise with these rhythms; it is only with some effort or through such projects as rhythmanalysis that we become critically aware of them. We can free ourselves from them, or reinvent them, only through the expenditure of the creative resources available to us.

Although social and political, Lefebvre's project to analyse the rhythms in social and cultural life, which become second nature through somatic learning, is fundamentally cross-disciplinary. The rhythmanalysis he advocates would benefit from being sensitised to biology, physics and psychology, which internalise and cement standards of social behaviour through a lifetime of learning the nuances of body posture, gesture and inflection appropriate for various situations (the shop, the school, the army, the bathroom). For Lefebvre, rhythm is a form of repetition that can produce difference and complications – polyrhythms, eurhythmia and arrhythmia – the observation of which can be grasped 'preconceptually', such as when we observe the sea with its superpositions and variations.[19] Lefebvre contrasts linear and cyclical principles of rhythm to arrive at a complex understanding:

Cyclical processes and movements, undulations, vibrations, returns and rotations are innumerable, from the microscopic to the astronomical, from molecules to galaxies, passing through the beatings of the heart, the blinking of the eyelids and breathing, the alternation of days and nights, months and seasons and so on. As for the linear, it designates any series of identical facts separated by long or short periods of time: the fall of a drop of water, the blows of a hammer, the noise of an engine, etc. . . . There is between them an antagonistic unity. They penetrate one another, but in an interminable struggle: sometimes compromise, sometimes disruption. However, there is between them an indissoluble unity: the repetitive tick-tock of the clock measures the cycle of hours and days, and vice versa. In

---

[19] Lefebvre writes:

Watching waves, you can easily observe what physicists call the superposition of small movements. Powerful waves crash upon one another, creating jets of spray; they disrupt one another noisily. Small undulations traverse each another, absorbing, fading, rather than crashing, into one another. Were there a current or a few solid objects animated by a movement of their own, you could have the intuition of what is a polyrhythmic field and even glimpse the relations between complex processes and trajectories, between bodies and waveforms (Lefebvre 2004, 79).

industrial practice, where the linear repetitive tends to predominate, the struggle is intense (Lefebvre 2004, 76).

The body becomes a metronome, with the intersection of internal and external rhythms complicating the exchange. Lefebvre anticipates in broad strokes the nonlinear dynamics of the brain and embodied processes and refers to 'the polyrhythmic body' composed of diverse rhythms 'in a perpetual interaction, in a doubtlessly "metastable" equilibrium' (Lefebvre 2004, 80). The body is the 'intersection of the biological, the physiological (nature) and the social ... where each of these levels, each of these dimensions, has its own specificity, therefore its space-time: its rhythm' (81). Ultimately, the body is also the site where rhythm can be turned into creativity or subjugation. For Lefebvre, rhythm as a broader social phenomenon can free us from the everyday or make us a slave to its machinic processes:

The body and its rhythms remain no less a resource of music: the site towards which creation returns through strange detours (jazz, etc.). Musical rhythm does not only sublimate the aesthetic and a rule of art: it has an ethical function. In its relation to the body, to time, to the work, it illustrates real (everyday) life. It purifies it in the acceptance of catharsis. Finally, and above all, it brings compensation for the miseries of everydayness, for its deficiencies and failures. Music integrates the functions, the values of rhythm (Lefebvre 2004, 66–67).

Lefebvre feels, along with Schopenhauer, that music can provide the opportunity, on occasion, to break up the will of the world. For Lefebvre, the predictable, repeat rhythms of the everyday that regulate our lives translate themselves into our own rhythms, absorbing not only the rhythms of work, education, the news and consumerism, but even our free time, with standardised modes and rituals of play, entertainment and seasonal holidays. Can artworks reveal, parody, complicate or disturb these rhythms?[20] For Deleuze, Bacon's paintings can. Examples of abstract art and *Informel*, as well as earth art, through their various struggles between composition and non-composition of materials, produce complications that upset the rhythms of the everyday that we have internalised. And it is here, while we consider artworks that take us outside our habitual involuntary thoughts and rhythms of the everyday, that

---

[20] This echoes Deleuze's observation:

For there is no other aesthetic problem than that of the insertion of art into everyday life. The more our daily life appears standardised, stereotyped and subject to an accelerated reproduction of objects of consumption, the more art must be injected into it in order to extract from it that little difference. ... Art thereby connects the tableau of cruelty with that of stupidity, and discovers underneath consumption a schizophrenic clattering of the jaws (Deleuze 2004, 293).

metastability emerges. If Heidegger was right that the work of art reveals the struggle between the world and the earth (as well as the tension between tacit and explicit being-in-the-world), we can add to this Lefebvre's insight that this is a rhythmic struggle, one that perhaps brings linear clock time in conflict with circular biological time and its nested rhythms.

This is of course quite different from the philosopher John Dewey's pragmatism in *Art as Experience* (1934). Dewey places great importance on rhythm in art and writes that 'underneath the rhythm of every art and of every work of art there lies, as a substratum in the depths of the subconsciousness, the basic pattern of the relations of the live creature to his environment' (Dewey 1934, 150). Rhythm is fundamental in brain, body and art and is the common medium through which they communicate and breathe together. In the chapter on the 'Natural History of Forms', Dewey describes the rhythms of the sun and moon, the seasons, heartbeats, breathing, sleeping and waking and discusses the rhythms involved in preparing for war and for agriculture. Rhythms mark humans' experience of nature and memorialise intense and special experiences. For Dewey, the perception of rhythm in paintings and sculpture is as important as rhythm in music and is primarily felt directly rather than codified. The artistic composition of lines and colours is not simply representational or semiotic but produces an aesthetic experience that emerges from and further elaborates this rhythmic substrate.

Although Dewey understands art as a special intensification of the human relationship to the environment, which is rhythmic, he insists that art emerges from everyday existence and should not be set above it on a pedestal. For Lefebvre, art must break the spell of the everyday, which leads us to drift into ritual rhythmic repeats that smother creativity. It is interesting that many artists and viewers of art struggle between these two poles. Think, for example, of how Andy Warhol's *Green Coca-Cola Bottles* (1962) might be understood as a 'transfiguration of the commonplace' using the commonplace, while Pollock's *Autumn Rhythm (Number 30)* (1950) is a transformation of the commonplace using extraordinary means.[21] In *Man's Rage for Chaos*, an influential book for many artists, Morse Peckham writes that the artist 'offers us the experience of disorientation' (Peckham 1965, 79). Peckham describes how he was utterly bewildered when he saw his first exhibition of abstract painting and then noticed a peculiar phenomenon where traditional painting 'began

---

[21] The phrase is taken from Danto (1983). The author argues that art makes the commonplace strange and that artistic theory is crucial in helping us to discern the distinction between the ordinary and art.

to look very different. It began to look in an odd way like abstract painting'
(Peckham 1965, 153). The artist Dubuffet similarly declared that art
should 'disorient you powerfully' and that when it 'loses the quality of
its strangeness ... it loses all efficacy' (Dubuffet et al. 2006, 147).

Lefebvre's rhythmanalysis is important in creating a paradigm shift in
how we perceive mind and matter, from a clear dualism towards
a complexity of rhythms, circular and linear, emerging from brain, body
and world. At a deeper level of analysis, Lefebvre describes rhythm as
a form of expenditure, a form of energy: 'An energy is employed, unfolds
in a time and a space (a space-time). Isn't all expenditure of energy
accomplished in accordance with a rhythm?' (Lefebvre 2004, 65).
Dewey was fond of using the terms 'energy' and 'exchanges of energy'
which the artwork was able to supply. For Lyotard, working in the same
historical context as Lefebvre, Serres and Deleuze, all of whom were
interested in scientific concepts to do with matter, the difference between
mind and matter is one of degree, and the continuity between mind and
matter 'appears as a particular case of the transformation of frequencies
into other frequencies, and this is what the transformation of energy
consists in' (Lyotard 1991, 43). He goes on to explain 'the mind' using
terms informed by science and, to some extent, anticipates the neuroscience of chaos in the brain:

Contemporary science, I believe, shows us that energy, in all its forms, is distributed in waves. .... The reality to be accorded to such-and-such a form of energy,
and therefore of matter, clearly depends on the transformers we have at our
disposal. Even the transformer that our central nervous system is highly sophisti-
cated in the order of living creatures, can only transcribe and inscribe according to
its own rhythm the excitations which come to it from the milieu in which it lives
(Lyotard 1991, 43).

These 'waves', 'vibrations', 'transformers' and 'rhythms' have their
scientific counterparts, which I examine in Part II.

For Michel Serres, inspired by Lucretius, the aleatory is chaotic matter
that is unpredictable and takes its own course.[22] During the straight
trajectory of falling, there may be a chance encounter between atoms
which 'swerve', setting a whole series of events into motion. The universe
has no 'purpose' or teleological progress but is predicated on chance, so to
embrace lived aleatory and unpredictable events and encounters is to move
towards freedom and chaos and creativity. In Serres, we arrive at the
junction between matter, chaos and psychology. If for him psychology is
another word for physics, it would not be Newtonian, and certainly not
linear. This physics of psychology, which anticipates the quantum physics

---

[22] For an introduction to Serres, see Abbas (2005).

of the brain, would be more aligned to the actions of mind wandering than rational induction, and this chimes in with recent studies of chaos in the brain. For Hanjo Berressem, in the work of both Serres and Deleuze 'the Lucretian clinamen is shorthand for the logic of nonlinear dynamics' (Berressem 2005, 51) which is also entirely relevant for matters of the mind:

> [T]he intensities of sudden, violent chance events are stored in various ways in the psychomaterial archives of the subject. They are 'impressed' into the landscapes of thought and writing (where these impressions – or, more precisely their flashes – are transferred, at least partly, into the landscape of conscious and unconscious memories), but they are also impressed into the neuronal landscape (often to such a degree that they can no longer be accessed except in the sudden flashes of what is not quite correctly called *mémoires involontaires*) and into what might be considered muscular and visceral archives – archives of the flesh – with their own mnemonic operations (Berressem 2005, 66).

The neuroscientist Antonio Damasio suggests that this memory – 'archives of the flesh' – does not just reside in the hippocampus but is embodied as a somatic marker. These markers could be pulse rate, heartbeat, or pain and aches in joints or muscles, triggering coordinated events in the nervous system, muscle memory and affects. There seems no reason why these somatic markers, situated in the biological system, cannot be activated by cues in films or artworks.

Providing another dimension to this discussion is Jessica Wiskus, interpreting French novelist Marcel Proust's *Remembrance of Things Past* (1913): '[I]t is not the *past* that inspires within Marcel a feeling of joy – it is the *rhythm* of the past and present' (Wiskus 2013, 120). One way of understanding this is to imagine a convergence of events that create a distinct set of rhythms: a butterfly, some flowers, a particular breeze stirring the leaves, creating a feeling of contentment. A somatic marker, the embodied feeling of contentment, is attached to a particular set of rhythms and variations of colour. And if, years later, the rhythms are combined and repeated, triggering a similar contentment (which is a form of rhythm appreciation, now accompanied by memory), it is not because the past or the present are pre-eminent so much as what is common to both: the rhythms and sensations felt by the body. This is of course articulated best by Proust himself when he writes of one of his characters, 'the flowers of the past are true because he has carried them along inside himself – or rather, not inside himself but as himself' (quoted in Wiskus 2013, 36).

This understanding of the archives of the flesh is consonant with Bachelard's interpretation of Lautréamont's prose which evokes a muscular lyricism. The kind of writing that reveals the material movements of vibration and rhythm, and the reveries that amplify them, Bachelard calls 'dreaming aloud' ([1948] 2002, 244). Daydreams, as we know, can occur while watching castles and elephants form in the clouds, or following veins in the sea, or roots threaded along the floor of forests. Guattari writes that 'delirious narrativity', what I would understand as nonlinear narrativity, which we might see in Bataille or Lautréamont, is a 'non-discursive substance [that] constitutes the paradigm for the construction and reconstruction of mythical, mystical, aesthetic, even scientific worlds' (Guattari 1995, 82). Rhythmically this is a kind polyphony, or sometimes even a cacophony, which allows many singularities and heterogeneities to come together illogically in the encounter with art. A visual example of this is collage, which, as with stuttering, tears apart and reconfigures conventionally understood continuities in form so that recognition is fractured, roughly joined, preposterous or unsettling. This breaking up of what should be known quantities and unities creates an eerie world with its own logic.

1.6   Anton Ehrenzweig: Daydreaming Matter

In *The Hidden Order of Art* (1967), Ehrenzweig combines aspects of psychoanalysis, art history and philosophy to develop insights into the creative imagination involved in the production and reception of art. In so doing, the work shares common approaches with Bachelard, Deleuze and Guattari, and modern psychologists who study creativity. A key passage in the book details Ehrenzweig's complex understanding of the internal and external mechanisms involved in observing art. Not only does he take the standard psychoanalytic line that non-conscious processes continue to run underneath waking experiences, he also maintains that images can sometimes alter the relationship of this 'underneath' activity to our conscious awareness. He writes that undifferentiated unconscious vision, in its

all-embracing sweep, can use almost any object form as an assembly point for an immense cluster of other images that, for a conscious analytic vision at least, have nothing in common. Any objects, however different in shape or outline, can become fully equated with each other on the unconscious 'syncretistic' level (Ehrenzweig [1967] 1993, 272).

Complicating the conscious/unconscious binary in traditional psychoanalysis are such concepts as the 'preattentive' or the non-conscious in psychology. Unbeknownst to us in our daily lives, our eye movements or saccades can amount to as many as 173,000 in a single day. It is striking and important that most of these go unnoticed by us. Vision relies on two cortical pathways: one mainly conscious (ventral stream including memory areas) and one non-conscious (dorsal stream). Every time we pick up a cup of coffee, or walk through a crowded cafe, or catch a ball, we are using non-conscious vision.

Kandel expresses this as a 'preattentive' process concerned only with the general or global features of the object, such as shape and texture (something that matter painting and abstraction actually foreground), followed by an attentive process of interpretation using the ventral stream, which involves various kinds of memory, as well as temporal lobe linguistic processing (Kandel 2016, 29). But even this formulation may be too starkly binary, as these processes cooperate, and works of art allow them to work together, drawing upon a possibly unique and certainly metastable duration of mental resources activated in pulses and oscillatory frequencies.

Ehrenzweig brings together the act of observing things in the external world and different states of consciousness. The first is *differentiated*: conscious, analytical, self-aware, with the broken rhythms of eye movements rapidly targeting shapes and forms to gather information. This is contrasted with the *undifferentiated* mode of observing, which is more sensory and less hyper-aware, and which may take the whole vista into account in its general outlines as a motion rather than as a set of points to be discerned. In more recent psychology, one might see the terms 'local focus' and 'global focus' used to describe this contrast. Importantly, Ehrenzweig links the ability to switch from one mode to the other with creativity:

The creative thinker is capable of alternating between differentiated and undifferentiated modes of thinking, harnessing them together to give him service for solving very definite tasks. The uncreative psychotic succumbs to the tension between conscious (differentiated) and unconscious (undifferentiated) modes of mental functioning. ... Modern art seems chaotic. But as time passes by the 'hidden order' in art's substructure (the work of unconscious form creation) rises to the surface. The modern artist may attack his own reason and single-track thought; but a new order is already in the making. Up to a point any truly creative work involves casting aside sharply crystallized modes of rational thought and image making (Ehrenzweig [1967] 1993, xiii–xvi).

Global, configural scanning is important for Ehrenzweig as a way of avoiding intellectual fixation on details. The scattering of focus is a way of causing de-differentiation. This is different from gestalt aesthetics that emphasises shape and scene recognition, based on prior learning of 'good form' or by picking out or differentiating details from the visual field. Related to this is the term 'configural'. Zangemeister et al. (1995) studied the eye behaviour of art experts and novices and found that those with little experience of viewing artworks tended to move their eyes across shorter distances, particularly when viewing abstract art, compared to those more acquainted with art who moved their eyes over the whole picture. This is also what artists tend to do (Nodine and Krupinski 2003), which suggests that configural viewing takes in the broader aspects of a composition. This coheres well with results presented by Cela-Conde et al. which show that experts process configural and global shapes and forms rather than fixating on particular details (Cela-Conde et al. 2011, 45). The expert is able to focus on particular fine-grained structures when required, while working memory keeps the global representation in mind. In Rudolph Arnheim's gestalt psychology, the mind naturally seeks integrated structures as a form of visual behaviour, and vision automatically fixes the centre and periphery, verticals and horizontals, and various kinds of shape recognition as 'perception centres', reading relationships between them: 'motifs like rising and falling, dominance and submission, weakness and strength, harmony and discord, struggle and conformance' (Arnheim 1971, 434). In other words, the organisation of complex concepts becomes analogous with the geometric arrangement of perceptual cues in the artwork. For Arnheim, the end result of art is to establish a pleasing order or balance of stresses – an aesthetic concept superimposed over a perceptual tendency identified as 'natural'. What need to be avoided at all costs are conditions of imbalance, where 'the artistic statement becomes incomprehensible. The ambiguous pattern allows no decision on which of the possible configurations is meant' (Arnheim 1971, 20). This is a strong argument for design and for explaining the tendencies of vision when observing the ordinary world, but does it identify a universal law of art?

For Ehrenzweig, it is *disintegration* that can equally be part of the creative process, as well as disorder. Chaos is not simply raw material used to service the pleasures of unity once resolved, a strategy favoured in traditional aesthetics and folk psychology. Even though gestalt psychology aimed to get away from reductionism in order to view the whole, Arnheim's over-determination of unity and order as the most important principles or aims of aesthetic experience blinded him to the great diversity of artistic practices which have other ends and which show us the discordant and chaotic aspects of human nature and the universe. Against the gestalt understanding of art's purpose, Ehrenzweig writes that 'we have to suppress our interest in pattern as such in order to make vision into an efficient instrument for scanning reality' (Ehrenzweig [1967] 1993, 14). For him, 'blurred plasticity is more important for the efficiency of vision than making out precise shapes and patterns' (15). Undifferentiated vision thus becomes a particular method of producing and experiencing form in art (more heavily emphasised as 'matter' in Smithson and Morris), which defers the identification of fixed geometries in favour of unfolding complexity, as we experience when our reverie alights upon the coals, intensifies the sparkling foam of the sea, or allows us to breathe with the swelling musculature of a soaring altocumulus. This reverie is not just in the mind; it requires light to be rhythmically filtered in certain ways to scatter rational thoughts.

The different levels of differentiation are apparent in a deep 'oceanic undifferentiation' and during de-differentiation: 'It is possible to train one's powers of introspection to hold onto the less articulated states of consciousness and to earlier phases in the history of perception where its gestalt structures were not yet fully crystallized' (87–88). This method of observing art requires patience and openness to fluctuations in the visual field by deferring gestalt structures that feed into traditional aesthetic learning concerning form and composition. A certain mindfulness is required to 'disperse' these structures, however, which is easier to do when the paintings themselves are organised to support this kind of de-differentiation, as we see in different kinds of matter painting.

This obviously has much in common with aspects of creativity involved in mind wandering, which may involve the deferral of prior meaning. Moshe Bar (2009) suggests that we generate several possible analogies, associations and predictions for a single input stimulus or problem, so that the mind has multiple interpretations of an event 'available'. We may bring any of these interpretations to the fore in consciousness, which suggests that they are held somewhere in less-than-conscious readiness, but this bringing to the fore may be triggered by salient material features or configurations in the environment. Perhaps art allows this process of delaying resolution so that it can be savoured. In well-known studies of metaphor, weakly intended meaning remains active during metaphor comprehension (Bowden and Jung-Beeman 2002). Similarly, in problem-solving, subjects maintain weak activation and remote associations for the solution, alternatives that can be called upon to achieve insights. In these examples it seems important that we entertain a wide openness to possibilities, suspending strong commitment to any of them. This feeds into so-called divergent thought, which is loosely or spontaneously composed and proceeds by making unusual far-flung connections. This is different from convergent thinking, which relies on well-known formulae.

This passive receptiveness, as opposed to active manipulation, occurs when one is being patient with apparently meaningless chance sequences. There are echoes here of Bergson's detension. This psychology of 'letting matter be' can also be understood in terms of Heidegger's notion of releasement or deferred willing (*Gelassenheit*). As Ehrenzweig further elucidates, in abstract expressionism

we have to give up our focusing tendency and our conscious need for integrating the colour patches into coherent patterns. We must allow our eye to drift without sense of time or direction, living always in the present moment. … If we succeed in evoking in ourselves such a purposeless daydream-like state, not only do we lose our sense of unease but the picture may suddenly transform itself and lose its appearance of haphazard construction and incoherence (Ehrenzweig [1967] 1993, 74).

For Heidegger, authentic releasement is obtained when willing (the desire to make something meaningful to us or useful for us, for example) has turned into a will-less state of being non-desiring, letting things be rather than forcing one's being over other beings. In this way, reverie can be seen as a will-less, passive watchfulness that follows the rhythms of the thing observed, allowing it to take its course, not cutting up its duration by glancing at a wristwatch and asking it to perform.[23]

Ehrenzweig's de-differentiation helps us to understand that mind wandering can also be seen as 'eye wandering', vision free from a driven goal, panning over details in a disorderly manner.[24] One is free to switch to a more rational and analytical mode of inspection, and sometimes one finds insights by re-entering this mode after mind wandering. It seems important that this switching itself has a rhythm; Ehrenzweig speaks of a rhythmic switching from one kind of observing art to another, moving from differentiation to de-differentiation.[25]

Among the higher-order concepts that can be associated with this disordered meandering, different from the aesthetic ideals of order and unity espoused by Arnheim, is the novelist James Joyce's 'chaosmos', which describes a coming together of disorder *and* order, existential freedom *and* rules, chaos *and* cosmos. In Joyce these concepts are not

---

[23] This is not just relevant as a way of trying to solve a problem or efficiently complete a task. Cultivating this openness is a way of seeing art. It requires patience before visual and mental connections are made that otherwise would be overlooked if in a hurry to execute the plan. This is also articulated in philosophy by Heidegger, who expresses an authentic kind of waiting (*Warten*): 'waiting on nothing' (Heidegger 2010, 140), which Bret Davis describes as

[an] attentive and engaged openness to an arrival of something unexpected, in contrast to an awaiting (*Erwarten*) that would first actively project what it then expects to passively receive. A genuine non-willing waiting would be neither a merely passive reception of a fate nor an aggressively active projection of a plan (Davis 2010, xv).

[24] An elegant study by Alfred Yarbus, which continues to influence eye tracking research today, allows us to be aware of the effects of priming subjects. Viewers can be 'primed' to adopt certain routines of eye movement: analytical or purposeful if they are asked to look for, or identify, certain features in a picture, which may override the formal or sensuous qualities. Withholding these instructions may lead to a more passive, arbitrary viewing, accepting rather than overriding the perceptual cues provided by the artist (Yarbus, 1967). Of course, it is possible to cultivate the latter kind of observation of art by trying to avoid keeping a narrow purpose in the mind, which more easily allows for reverie, as opposed to extracting information and matching it to prior purpose, which leads to an entirely different feeling or experience.

[25] Ehrenzweig notes:

As we penetrate into deeper levels of awareness, into dream, reveries, subliminal imagery and the dreamlike visions of the creative state, our perception becomes more fluid and flexible. It widens its focus to comprehend the most far-flung structures. These different levels of differentiation in our perception interact constantly, not only during the massive shifts between dreaming and waking, but also in the rapid pulse of differentiation and de-differentiation that goes on undetected in our daily lives (Ehrenzweig [1967] 1993, 87).

opposites but part of a larger continuum. With the psychology involved in making art, one has to find a balance between conscious control (order) and free play (chaos) that is not so controlled, and one gets better at producing (and being able to recognise) this chaosmos with practice.[26] Thus the psychology that produces chaosmic works, which I believe matter paintings make manifest, is itself a form of chaosmos, and when viewers observe the outward marks of this in matter painting, they too may enter a psychology of chaosmos. Another, more technical term for chaosmos is 'metastability'. It seems important that this metastability, when viewed configurally rather than piecemeal, is not just in the head but is produced by brain oscillations (neuroscience), sensations (phenomenology) and composition/non-composition (art).

In Joyce, the style of writing draws attention to textures, to oneiric and visceral processes and rhythms that carry striking and irrational juxtapositions and fragments of images. Ehrenzweig describes Joyce, Cubism and matter painting ('tachism') in these terms:

Bion [the British psychologist William Bion 1897–1979] has said that schizoid splintering of the language function does not prevent a creative use of language if unconscious linkages are preserved. James Joyce's splinter language is of this kind. His phantastic word conglomerates are not just violent compressions of language splinters, but establish counterpoints of dreamlike phantasies that run on below the surface and link the word clusters into an unending hypnotic stream. Similarly the fragmented, violently condensed picture plane of a ripe Cubist painting is held and animated by a dynamic pulse. It draws the fragments together into a loose yet tough cocoon that draws the spectator into itself. Again this space experience does not lack a hypnotic, almost mystic quality. The new American painting has ground down all splinters into tachist shreds and textured fragments. I have mentioned how it represents the climax of a long development of splintering that probably began already with French Impressionism (Ehrenzweig [1967] 1993, 118).

Many literary works attempt to juxtapose unusual ideas or images and enhance this striking effect with intervals, cuts and overlapping durations. Writing in the nineteenth century, psychologist William James explains this in the following manner:

Instead of thoughts of concrete things patiently following one another in a beaten track of habitual suggestion, we have the most abrupt cross-cuts and transitions

---

[26] An important characteristic of the thought involved in producing remote associations is nonlinearity, where spreading activation is not controlled by logical sequences. Related to this is recent research into the 'cognitive shuffle' or 'serial diverse mentation', which involves imagining random objects in order to trigger the onset of sleep, easing out of worrying or ruminating which would keep us awake. The fact that the random selection of objects in remote association can be used to spur poetic insight as well as to fall asleep suggests that both processes are related to the brain's resting state where drowsy mind wandering occurs (Burkeman 2016).

from one idea to another ... the most unheard-of combinations of elements, the subtlest associations of analogy; in a word, we seem suddenly introduced into a seething caldron of ideas ... where partnerships can be joined or loosened in an instant, treadmill routine is unknown, and the unexpected seems the only law (quoted in Becker 1995, 222).

In music, Beethoven, in later years, preferred his compositions to be more chaotic, which Ehrenzweig describes as a 'dispersed unfocused polyphony' (Ehrenzweig [1967] 1993, 89).[27] And, of course, jazz improvisation is another example of 'chaosmic' composition. Ehrenzweig mentions 'tachism' (a name given to post-war French artists using dabs and splodges), where this chaosmic aspect emerges in artistic practices that welcome 'little accidents' – unplanned textures and blends in tone or colour that allow unexpected innovations or design solutions to occur. Mind wandering brings together splinters, odd juxtapositions and leaps of thought, as described by William James, emphasised by Ehrenzweig as a way of seeing and making art. This intersection of different perspectives helps us to understand the fine-grained nature of how ordered conscious thought can be interspersed with improvisation. On this view, mind wandering should not be seen as a turning inwards and away from the external details of artworks, but as a special way of engaging with them. As creative rhythmic thought, mind wandering allows us to latch onto the formless and arbitrary aspects of artworks, often preserved in order to facilitate this very experience for the viewer. Seeing in this way, the physical trace of what the matter suggested to the artist can be imagined by the viewer. Such spells would only be broken by 'making sense' or insisting on a linear narrative.

This way of allowing oneself to be open to or anchored in the inanimate has a long tradition in art practice. Perhaps one of the most celebrated examples of this kind of 'viewing matter' is Leonardo da Vinci's advice to students to observe the arbitrary stains and irregularities on the wall. In the scraps of text we have come to know as his *Treatise on Painting* (1270), Leonardo explains:

Do not despise my advice, which reminds you ... to stop sometimes and look into the stains of walls, or the ashes of a fire, or clouds, or mud, or other like spots, in which, if you consider them well, you will find the most marvelous inventions; for the *ingegno* [imaginative ingenuity] of the painter is roused to new inventions,

---

[27] Ehrenzweig discusses how Beethoven

took a disruptive idea which guides and unfolds the large-scale structures. A fully articulate well-knit melody belongs too much to consciousness. An incoherent fragment, a disruptive form element is better able to break the narrow focus of intellectual thought and produce a fissure in the mind's smooth surface which leads down to the depth of the unconscious (Ehrenzweig [1967] 1993, 50).

whether of compositions of battles of animals and men, or of various compositions of landscapes and monstrous things ... because the *ingegno* is roused to new inventions by indistinct things [*cose confuse*] (da Vinci [1270] 1956, 37).

In the modern era, we have the example of André Masson's well-known autonomist drawing *Shadows/Les ombres* (1927), which appears to be primordial writing in the sand, outlining random shapes and forms (Figure 1.3). It seems as if Masson is following Leonardo's advice, but using sand instead of stains on the wall, while also, remarkably, managing to preserve something of the original relinquishment of control, so that the viewer may exercise their imagination on the wandering line by following a similar capriciousness. We might imagine ourselves, like the artist, in a liminal zone between matter 'for itself', as arbitrary marks, and matter 'for us', the shifting shapes we dimly discern and which seem to self organise in the visual field. Such an experience may even act as a kind of therapy, where the line changes its course depending on whether memory, perception, past or present is uppermost in the mind. The artist

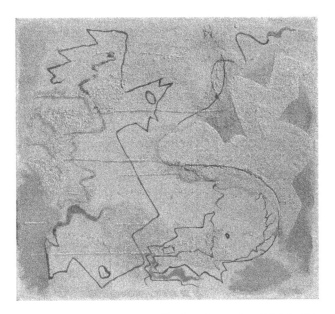

Figure 1.3 André Masson, *Shadows/Les ombres* (1927). Oil and sand on canvas, 50.2 × 54 cm. Digital Image © Christies/Scala, Florence © André Masson/ADAGP. Copyright Agency, 2020. (A black and white version of this figure will appear in some formats. For the colour version, please refer to the plate section.)

or designer's *ingegno*, exercised by an engagement with the relatively formless, is intentional yet also somewhat relaxed about unintended occurrences. In this sense, the line in the drawing does not just represent an object or thing but actually traces the tension between willing and non-willing, representation or formal structure and de-differentiation, and the viewer shares the experience of this tension with their own meandering between these states.

This was grasped by the Surrealist André Breton, who suggested that automatist writing achieved 'the relaxation of emotional tensions due to repression, a lack of sense of time, and the replacement of external reality by a psychic reality obeying the pleasure principle alone' (Breton 2002, 70). In Smithson, influenced by Ehrenzweig, we have a clear acknowledgement of the kind of psychology required to produce art, which, at least for the artist, meant trying to shrug off the inhibitions of reasoning:

At the low levels of consciousness the artist experiences undifferentiated or unbounded methods of procedure that break with the focused limits of rational technique. Here tools are undifferentiated from the material they operate on, or they seem to sink back into their primordial condition. The rational critic of art cannot risk this abandonment into 'oceanic' undifferentiation, he can only deal with the limits that come after this plunge into such a world of non-containment (Smithson 1996, 102).

'Sinking back' to a 'primordial condition' of 'non-containment' and undifferentiation are concepts strongly influenced by Heidegger, and they point to the fundamental attunement of some kinds of psychology with philosophy, insofar as undifferentiation, which we may achieve in moments of reverie and mind wandering, has both an epistemology as well as ontological facets. Forms of abstract art can work with us to sustain mind wandering as a kind of shape-shifting with indistinct matter. This finds a correspondence with the French critic Jean Paulhan's thoughts about *Informel*'s metamorphoses, which can

transform a mouse into an elephant or pumpkin into a carriage ... the objects metamorphosed belong to two different orders: one to our inner and personal thoughts, the other on the contrary to external objects ... such metamorphosis does not take place in a fairy tale; nor does it exist for credulous children. Serious minds, trained in scientific methods: Langevin, Sartre propose this to us very naturally. As though we were so profoundly mixed up with the world and so merged with it that our most personal thoughts were less thoughts than reality (Paulhan 2002, 186).

This is a kind of visual stream of consciousness where forms continually bubble up and disperse. Conscious control and volition drive a route through such 'noise' in pursuit of our daily tasks. But, as the poet Percy Bysshe Shelley notes, 'poetry is not like reasoning, a power to be exerted according to the determination of the will. A man cannot say "I will compose poetry"' (quoted in Wallas 1945, 75–76). It is important to state that a work of criticism or art history may proceed by rational analytical means, but this procedure does not reflect the contents of the thing analysed or what it feels like to be immersed in this 'interval free from conscious thought' (Wallas 1945, 46).

Mark Runco, one of the foremost experts in the scientific study of creativity, writes:

[T]he unconscious is less prone to censoring, and as such it has a higher likelihood of generating remote associates and original ideas. Another way of describing the benefit is that the use of pre-conscious or unconscious processes allows the individual to utilize different reasoning processes, processes that, by virtue of their being beyond conscious awareness, are able to value and explore those things that allow original thinking. In this light the preconscious and unconscious are not actually irrational; they just have a rationality of their own (Runco 2006, 109).

And it is Ehrenzweig, again, who understands this point well in his description of the earlier stages of learning to create art:

One has to explain to the student that a purely conscious control of the working process is neither desirable nor possible. The rigid art student often comes to the college motivated by the pious wish for good craftsmanship, which he misconceives as a fully conscious control of his medium. He wants to put down 'what he has in mind'. It is useless to explain to him that what he has in mind are usually clichés and mannerisms which he has picked up from existing art in a life-long devotion to the master works of the past . . . . True craftsmanship does not impose its will on the medium, but explores its varying responses in the kind of conversation between equals [which consists in a] passive but acute watchfulness for subtle variations in the medium's response (Ehrenzweig [1967] 1993, 57).

This advice is equally important for the viewer, who also seeks consciously to control their engagement with art and to impose on it mannerisms of the past. Particularly with abstract art, a 'passive but acute watchfulness' allows the mind to wander with the artwork and follow the patterns of matter until some insight seems to occur of its own accord.

With mind wandering one is open to non-rational, nonlinear and relatively chaotic expositions of thoughts and mental images engaging with the rhythms of matter in art. Although art requires voluntary focus and attentive viewing, it also encourages play. This involves not imposing the will on matter or the artwork. The problem with intention is that it is too often associated with the immediate assertion of will, when in fact the will is dissolved when listening to a teacher or learning from events, so intention is often sacrificed in order for insight to occur. We say that a patient or student internalises the analyst or teacher so that they can help themselves. In a similar fashion, we can learn from a series of events, observing stains on a wall, as Leonardo suggests, to perceive anew complexities in form and pattern – we 'internalise' the stain as 'teacher'. This disturbs the long-standing tradition in philosophy that the mind and form act upon passive matter. When learning how to cultivate mind wandering as a way of engaging with a work of art, we become sensitive to the rhythms that it suggests. Rational calculation turns away from these rhythmic connections to follow its own sequences. Paradoxically, feeling rhythmically connected, as opposed to rationally attentive, produces insight.

Part II
Rhythms of the Brain and Matter Outside of It

Part II examines studies of metastable rhythms in the brain, particularly the rhythms involved in mind wandering, sustained by the brain, body and art. I draw upon empirical studies which reveal how the brain functions as a system of numerous unstable networks, where neurons are jittering on the edge of chaos, continuously ready for and acting in concert with 'perturbations': unanticipated abstract patterns and rhythms in the external environment.

2.1   Buzsáki

The rhythms described in the previous section find their technical and empirical counterpart in the dynamics of neuroscience. György Buzsáki is one of the leading neuroscientists specialising in neuronal oscillations and their complex overlapping relationships. In *Rhythms of the Brain* (2006), Buzsáki begins by describing the structure of neuron types and their specialised functions. Some neurons respond to highly specific stimuli, such as neurons specialised in responding to visual stimuli which detect movement, light, colour, lines and edges. There are many different kinds of neurons specialising in many different functions. When a neuron responds to a stimulus, it transduces this input into an electrical current or pulse which can act as a signal. Buzsáki describes the different ways in which a neuron can behave: it does not just produce a pulse but can also remain relatively inactive (not producing a pulse), act as an inhibitor, be entrained by (made to follow) a group of neurons, or contribute to the entrainment of other neurons. A combination of pulses from different neurons may become entrained or synchronous, vibrating together, creating fast, medium or slow frequencies or rhythms. The combined pulses of neural populations are wave-like in nature and are therefore described in terms of frequency. A frequency (or band) describes the number of pulses per second (hertz) of such synchrony. An electroencephalogram (EEG) detects electrical pulses in brain areas, measuring their frequency.

The brain is made up of billions of neurons creating uncountable numbers of interactions and functions via electrical pulses, the rhythms of which are identified by their frequencies. Switches in these frequencies depend on whether the subject is sleeping, mind wandering, listening to music or actively engaged in problem-solving. The major frequency bands have been identified as: (1) beta (14–40 Hz) associated with the normal waking state of alertness and reasoning; (2) alpha (7.5–14 Hz) observable in states of relaxation but also in daydreaming, mind wandering, meditation and incubation of ideas; (3) theta (4–7.5 Hz) for light sleeping including REM periods but also present during deep meditation; and (4) the deep sleep wave, delta (0.5–4 Hz). There is also a very fast frequency gamma wave (above 40 Hz) present in moments of insight, rapid thought processes and high-level information processing. It is important to remember that these frequencies often work across brain networks, rhythmically entangling, for example, the functioning of the default mode network (DMN), which supports mind wandering and involuntary thought, or the executive control network, which supports focused attention and control. This provides some of the neural-level description of the phenomenal richness of the rhythms and durations of thought activated in viewing artworks. Particularly relevant is the question of how theta and alpha band activity, typical in mind wandering and meditation, is important for insight of the kind that is often reported in encounters with art. A complex picture emerges of a simultaneity and interaction of neural events on many levels from micro to macro, and nonlinear dynamics helps us to envisage this complexity.

According to Buzsáki, within the brain 'the several oscillatory classes have distinct mechanisms, each serves a different function. . . . Because many of these oscillators are active simultaneously, we can conclude that the brain operates at multiple time scales' (Buzsáki 2006, 117). There is also a complex system of 'nested' frequencies, which adds to our understanding of the simultaneity of rhythms. Some of the slower waves, for example, can act as 'carrier waves' for faster cycles: 'a theta wave can host between seven and nine gamma cycles' (352) and, importantly for being able to keep in mind items on a list or multiple features of an abstract painting, 'nested gamma waves can be used to simultaneously maintain several items in working memory in humans. The theta period would define the span of memory with seven to nine items multiplexed on successive gamma cycles' (352). Alpha waves strongly influence eye movements and are also attenuated by them: they can be 'blocked by various manipulations, such as eye opening, eye movement, visual imaginary and even mental activity, such as arithmetic calculations' (198). This is a remarkable example of how our bodies and the rhythms of the brain, in conjunction with what we are looking at (such as a work of art), can work together in a nonlinear dynamic way.

The study of brain frequencies from a dynamic systems or complex systems perspective runs counter to the popular image of the brain as a collection of regions with fixed functions and leads to a more complex understanding of the brain as a dynamic system of neural oscillations that travel across regions, making brain functions richly intertwined and in flux.[1] Studies of metastability in the brain emphasise that 'its dynamic ensembles [are] ever rearranging themselves as processes unfold that weave immediate and past events at numerous temporal and spatial scales' (Tognoli and Kelso 2014, 36). A change in dominant brain frequencies can occur when stimulated by external perturbations that can switch the coordination dynamics, particularly if the dynamics are poised 'at the border between order and disorder' (38). For Buzsáki, temporal coordination of synchrony 'is brought about by inputs from the physical world, much like the conductor's influence on the members of a symphony' (Buzsáki 2006, 154). He identifies how brain waves function in pulses and cyclically and how they entangle each other to create broader cycles of activity. Buzsáki describes this in simple terms; in recognising a face, for example, one may recall the person's

first and last names, her profession, our last meeting, and our common friends [which] are events that do not occur simultaneously but are protracted in time, since larger and larger neuronal loops must become engaged in the process. ... Each oscillatory cycle is a temporal processing window, signaling the beginning and termination of the encoded or transferred messages ... the brain does not operate continuously but discontiguously, using temporal packages or quanta ... different frequencies favor different types of connections and different levels of computation. *In general, slow oscillators can involve many neurons in large brain areas, whereas the short time windows of fast oscillators facilitate local integration* (Buzsáki 2006, 115–116, my emphasis).

This suggests a system of pulsing cycles working simultaneously. It is interesting that slow bandwidths such as alpha are active in daydreaming and mind wandering and travel across a larger brain area than faster bandwidths. This explains how divergent thinking, which employs the slow bandwidths, recruits a diversity of mental resources as opposed to the more constricted convergent thinking focused on a single problem.

One of the main effects of viewing and creating art is to strengthen the imagination. Art does this by providing images that help to produce

[1] Note that a particular oscillation can, however, function differently depending on the brain system that supports it; a gamma oscillation in the olfactory bulb is different from those in prefrontal circuits (Buzsáki, 2019: 157).

perceptual multistability and by prompting divergent thinking involving multiple conceptual associations. This may occur in cooperation with convergent episodes (focused, willed activity or conscious memorising) which aid evaluation of the broadband, often will-less mind wandering. This enables the retrieval of what is experienced during mind wandering, ensuring that it is not lost to consciousness, and this may be instrumental for insight (Ellamil et al. 2012; Dixon et al. 2014).

We are often not conscious of the way in which images encourage us to zoom in and zoom out. Also unnoticed by us is the way we rhythmically switch focus and return to areas of interest in cycles, which suggests the cooperation of broadband and narrow-band activity in metastable relationships. While we use our eyes in this way, co-present are shifting feelings, sensations and concepts that affect and are affected by this eye behaviour. The phrase 'affect and are affected by' already supposes that the brain, body and world share circular causality, with any one of them able to take the lead in shifting the predominant rhythms and dynamics of the art experience.

Art history can benefit from knowledge of these nonlinear dynamics. Developing a mathematically rigorous dynamic systems theory that traces one pattern of metastable behaviour across different domains or scales (micro to macro) is a hugely ambitious research objective even for science. Art history can, however, be sensitised to the rhythmic and iterative shifts simultaneously occurring at different levels of description. In this section, I will concentrate on four levels.

1. Metastability. Many neuroscientific studies have concluded that the dynamic properties of brain waves working together involve metastability and self-organising activity, often characterised as 'chaos in the brain'. This is important for understanding our engagement with external contents and environmental perturbations, which can act as switches for internal metastable dynamics. I examine the current neuroscientific research that helps to describe this interweaving of processes across brain-body-world.

2. Mind wandering. The slower frequency theta and delta bands in the brain's resting state sustain daydreaming, fantasies, memories and creative incubation. These frequencies are also observed during meditation. The current literature on mind wandering and the default mode network (DMN) is important, but the role of the DMN has not been examined as a key player in experiencing and producing art, which is often understood, particularly in neuroaesthetics, as predominantly the exercise of voluntary thought.

3. Perceptual multistability. The high-frequency gamma wave is associated with 'perceptual binding', which involves producing a unified

perception of different features of a scene or object, while the slower frequency involved in mind wandering creates 'perceptual multistability' (Van Leeuwen and Smit 2012), where shapes in the visual field can be seen in a variety of ways, sometimes in a fine-grained manner (gamma) or sometimes configurally and generally (delta). These different ways of seeing or adjusting focus can be explained as the cooperation of brain frequencies. The perceptual multistability produced by many artworks cooperates with phenomenal and conceptual dynamics. Perceptual multistability emerges when visual, auditory or olfactory patterns are too ambiguous for sensory systems to recognise in terms of one unique interpretation. Multistability involves transitions from one interpretation to another, driven by spontaneous and stochastic or chaotic neural events interacting with external cues. Metastability is a more complex term, taken from nonlinear dynamics, that describes how neural populations work by negotiating aspects of stability and instability. Thus, the experiential qualities of perceptual *multistability* are supported by aspects of neural *metastability*.[2]

4. Emotional processes. These accompany (1), (2) and (3) and involve an even broader metastability, along with the mind wandering and reverie cued by art. Marc Lewis (2005) explains emotional experience as the propagation of synchronised patterns of neural activity, which produces stability across different brain systems: brain waves propagate not only across particular brain areas but also across different brain networks, which are groups of brain areas working together. This rich and complex duration of emotional nuances, which emerges in particular social contexts, can be partly explained by how brain waves allow for cyclical, nested and complex variations and simultaneities.

The discussion and research gathered here concerning these four levels inform an art historical approach for describing how the rhythms of matter in art are keyed into the broader dynamic connectivity between brain, body and artwork. Artists and viewers intuitively gravitate towards the meditative, mind-wandering effects sustained by lower frequency bands even though they may know nothing of such brain frequencies. They experience reverie while making or observing art, in touch with or close to the qualities of matter, rather than rationally dissecting it, and this encourages a relaxed, non-rational, creative engagement that is important for resetting perceptual paradigms.

---

[2] Whether this can be described as neurofeedback is an interesting idea. See Bagdasaryan and Quyen (2013), which provides a theoretical justification for relating the variability of first-person experience to the variability of neural network dynamics, which they call 'neurofeedback'.

Neural frequencies working together in a metastable manner support the metastable aspects of conscious or semi-conscious experience that we see in mind wandering, kinaesthetics and motor imagery. Given that brain frequencies can act as carriers for other brain frequencies in nested formation, one might assume this also means that within a larger purpose or function such as a perception space (sustained by the carrier wave) there are diverse smaller tasks and functions that enter and leave the stage. This process can be explained by Leonard Talmy's example of how one perceives a dent in a car: it could entail a conceptual understanding that this is a damaged surface and not its perfect form; the perfect form might be a schema held in place in the working memory, perhaps with the aid of particular brain waves in the hippocampus. We could also run a mental script of cause and effect, the reason why or how it could have happened, with the addition of frequencies in other brain areas, or we might estimate the angle or force of impact, how expensive it might be to repair, and so on. This would all be happening within the 'perceptual space' but many of the cogitations are abstract. There are also temporal projections being processed while our sensory experience of visually attending to the dent continues. We have a grasp of the structural history or future of the object superimposed on the 'veridically seen static representation of the entity' (Talmy 1996, 255).

Although we are not aware of it, the multiplex operation of brain waves sustains this dynamic complexity at the phenomenal level. At the same time, the phenomenal complexity afforded by the dent in the car (or the artwork) helps to drive such polyrhythmic complexity. This may be an interesting complication in, or extension of, what neuroscientists refer to when they point to circular causality. Dynamic systems approaches allow us to understand that patterns or cycles of neural activity are repeated and sustained at different levels of organisation. This is a broad metastability which is 'chaosmic':

Order in the brain does not emerge from disorder. Instead, transient order emerges from halfway between order and disorder from the territory of complexity. The dynamics in the cerebral cortex constantly alternate between the most complex metastable state and the highly predictable oscillatory state (Buzsáki 2006, 135).

Importantly, this metastability provides us with our openness to the world:

Most times, brain dynamics are in a state of 'self-organized criticality.' This mathematically defined complex state is at the border between predictable periodic behavior and unpredictable chaos. In the context of brain dynamics, the implication of the concept of self-organized criticality is that the cerebral cortex displays perpetual state transitions, dynamics that favor reacting to inputs quickly and flexibly. This metastability is a clear advantage for the cerebral cortex since it can respond and reorganize its dynamics in response to the smallest and weakest perturbations (Buzsáki 2006, 128).

It is important not only that the brain responds to unpredictable environmental perturbations, but also that it does so by constantly shifting the emphasis of how it synchronises neuronal populations. Buzsáki acknowledges 'two sources of synchronization, externally driven and self-generated', which have 'tuning roles' (Buzsáki 2006, 154). Temporal coordination of synchrony 'is brought about by inputs from the physical world, much like the conductor's influence on the members of a symphony' (154). This suggests that a painting can act as a conductor. In extending the internal dynamics of the brain to external perturbations, the neurodynamics become more complex:

There are only two sources that control the firing patterns of a neuron at any time: an input from outside the brain and self-organized activity. These two sources of synchronization forces often compete with each other. If cognition derives from the brain, this self-organized activity is its most likely source. Ensemble synchrony of neurons should therefore reflect the combination of some selected physical features of the world and the brain's interpretation of those features (Buzsáki 2006, 155).

Buzsáki observes that neural oscillations can become synchronised with the rhythms of phenomenal experience, as we see with music and the visual rhythms of some artworks and films. However, the brain is also insulated from external 'perturbations', creating its own complex polyrhythms. Certain rhythmic perturbations in the environment can trigger cascades of neural populations, producing frequencies and rhythms that are not identical copies of the external perturbations that triggered them. In this sense, it might be more accurate to describe the effects on internal dynamics of what happens externally as resonances or translations. But there is also ample evidence of synchrony and entrainment, which suggests that a better word to describe these effects is 'participation'. A more technical term would be 'system coupling', where systems feature complex oscillatory behaviour leading to synchrony and asynchrony.

According to Buzsáki, certain kinds of meditation increase both the power and the spatial extent of alpha oscillations, depending on the expertise of the meditator:

Beginners show increases of alpha power activity over the occipital area, whereas in intermediate meditators the extent of oscillating cortical area is increased and the frequency of alpha oscillations is decreased. After decades of training, large-amplitude theta-frequency rhythm may dominate over a large extent of the scalp. Yoga and Zen training, therefore, reflects a competition between internal forces of synchrony and external perturbation .... The reasoning is this: when our brains are dominated by alpha oscillations, we feel a sense of calm; therefore, increasing alpha activity will calm our agitated brain. The repeatedly observed correlation between various laboratory measures of spontaneous alpha oscillations and cognitive and memory performance is a further rationale (Buzsáki 2006, 215–216).

Meditation reflects a competition between 'internal forces of oscillatory synchrony and external perturbation or changing spontaneous brain activity' (Buzsáki 2006, 216). Kelso et al. are in no doubt that 'while the anatomical structure constrains the dynamics, the dynamics simultaneously shapes the structure' (Kelso et al. 2013, 127). Tognoli and Kelso maintain that the complex coordination of multiple levels not only comes together to produce the richness of experience but should also be the means by which we explain such experience:

> We expect the coordinating mechanism of metastability to be realized at multiple levels (from molecules to minds) and in multiple functions ... for the metastable brain, active, dynamic processes like 'perceiving,' 'attending,' 'remembering,' and 'deciding' are not restricted to particular brain locations but rather emerge as dynamic patterns of interaction among widely distributed neural ensembles and in general between human beings and their worlds. Metastability offers scientific grounds for how cognitive processes come and go fluidly (Tognoli and Kelso 2014, 44).

The macro level of the phenomenology of perceptual multistability affects, and is affected by, the metastability of the micro level of neurodynamics. The artist's dynamic patterns of metastability propagate across various levels from neurological to sensorimotor and phenomenal to produce the artwork, which provides a way for us to become attuned to our own metastable dynamics.[3]

Mark Rothko's paintings in the famous chapel in Houston, Texas, have often been described in terms of their power to induce meditative states. Observing Rothko's dampening field, there seems to be few opportunities to count or discern objects that might focus the mind. The way our eyes adjust to subtle fluctuations of tones within a scale much larger than the human body seems to slow down time while expanding it (and expanding perception) into a smooth undifferentiated space. There is a sense of

---

[3] It is interesting that at the level of neural oscillatory frequencies, delta (broad and low) and gamma/beta (focused and high) frequencies work together. This suggests tendencies of chaos and order, where unfocused and broadly adventitious activity (itinerancy), and focused purposeful activity, come together in creative exploration, which can be viewed as a metastable openness to environmental perturbations. Tognoli and Kelso (2014) examine how the integrative tendencies of neuronal groups (suggesting activity towards the pole of order) coexist with segregative tendencies (towards chaos) to sustain a metastable state in readiness for a wide number of potential responses and transitions between them.

discovery in observing the treatment and application of the paint, which becomes more apparent the more one looks. Sylvester writes that some American abstract expressionist works have 'no major variations in density, changes of tempo' (Sylvester 1997, 66). There is an even-handedness in applying the paint, like the continual modulation of a machine buffing or sanding the floor without favouring one area over another, a delicate hand hovering over everything. But there is also patience, an even temper, where method is followed with discipline and poise. Buzsáki's observation that external cues can produce a sense of calm, which in turn 'further slows down micro-physical oscillations into more prolonged alpha activity' (Buzsáki 2006, 215), describes a feedback loop or circular causality where matter outside the brain and body becomes attuned to matter inside, to achieve what in philosophy might be called non-dualism.

A similar kind of circular causality occurs in the way our conscious thoughts process an artwork. Although the last thing you would want to think about in front of a Rothko is what your neural assemblages are doing, your thoughts might just wander to such matters of their own accord. This idea complements the notion of 'neurofeedback'. One may think of this as a contamination of the 'pure art experience', whatever that might be, but mind wandering and rhythmic sensations, which Rothko's works produce, alternate with crystal-clear aspects of linear and analytical thought or self-reflexivity.[4] Wang et al. (2015) suggest that viewers become more 'inward oriented' while viewing traditional Chinese landscape paintings.[5] They define 'inward oriented' as having higher levels of relaxation and mind wandering. Viewing realistic oil paintings has the opposite effect, where the viewer is occupied with object identification and attentive processes. The authors note that 'Chinese landscape paintings induce greater levels of relaxation and mind wandering and lower

---

[4] Lutz and Thompson clearly articulate that deeper knowledge of neurophenomenological processes does not have to interfere with the quality of the experience but, on the contrary, may train and reshape it:

One might therefore object that one form of experience is replacing another, and hence the new experience cannot be used to provide insight into the earlier form of untrained experience. This inference, however, does not follow. There is not necessarily any inconsistency between altering or transforming experience (in the way envisaged) and gaining insight into experience through such transformation. If there were, then one would have to conclude that no process of cognitive or emotional development can provide insight into experience before the period of such development. Such a view is extreme and unreasonable. The problem with the objection is its assumption that experience is a static given, rather than dynamic, plastic and developmental (Lutz and Thompson 2003, 48).

[5] But see also Starr (2013). Starr emphasises the importance of the self in the default mode network, whereas I emphasise the kind of mind wandering in the aesthetic experience where self-monitoring is reduced.

levels of object-oriented absorption and recognition, compared to realistic oil landscape paintings' (Wang et al. 2015, 327).[6]

Neuroscientist Walter J. Freeman writes that many techniques such as chanting, drumming and dancing – and, I would add, watching certain films or viewing abstract art – are ways of relaxing cognitive control to become sensitised to the macro and environmental context. Neuromodulators in the brainstem, such as oxytocin and vasopressin, further mediate this kind of chaos. It is significant that '[t]he world has effects on aggregates of microscopic neurons in the sensory organs. The brain acts into the world by microscopic motor neurons that send action potentials to muscle cells' (Freeman 2001, 21).[7]

Will and Turow have studied the entrainment of rhythms and synchrony across the domains of brain, body and world. They describe entrainment as a 'tendency for rhythmic processes or oscillations to adjust in order to match other rhythms' (Will and Turow 2011, 6), as we see with 'fireflies illuminating in synchrony … [or] human individuals adjusting their speech rhythms to match each other in conversation' (6). Tsuda et al. (2015) lend empirical support to this idea. They performed an experiment that alternated speech tasks between a computer and a human subject, and between two human subjects, in order to assess rhythm synchronisation. Human-to-human interactions produced positive subjective evaluations and entrained speech rhythms. EEG analyses demonstrated a significant increase in the amplitude of theta/alpha (6–12 Hz) rhythms compared to human–machine interactions. These results suggest that 'inter-brain synchronization of the theta/alpha rhythm … might reflect empathy for the rhythms of others during spontaneous temporal coordination' (Tsuda et al. 2015, 54). Strangely, we also see this in 'the movement of electric driers placed in close proximity … [and in] the way that a room of clapping people will spontaneously fall into rhythm' (54). Importantly, Will and Turow make a distinction: synchrony could be autonomous, involving separate non-interacting processes with their own energy sources, or it could be caused by the interaction of these processes. They point out that

[m]ere observation of synchronised behavior does not necessarily imply entrainment … . The wide range of entrainment phenomena is not based on a single physical process. Rather, the concept of entrainment describes a shared tendency of a wide range of physical and biological systems: namely, the

---

[6] In addition, Vartanian (2018) points to fMRI data that indicate activation of mind wandering in subjects viewing abstract art.

[7] The idea that the brain 'acts into' the world conveys Freeman's conviction that the brain does not merely think about the world but actively constructs it and interacts with it through the body.

coordination of temporally structured events through interaction. .... The interactions between oscillators can be unidirectional as, for example, in the case of human entrainment to diurnal cycles. Here the entraining oscillator (diurnal cycle) is not affected by the entrained oscillator (human being). However, in most cases of entrainment between living beings we are dealing with bidirectional or multi-directional interactions (Will and Turow 2011, 12).

The authors also mention circadian rhythms, heartbeats, blood circulation, respiration, locomotion, eyes blinking, hormone secretion and menstrual cycles:

It has been suggested that all human movements are inherently rhythmic and these endogenous rhythmic processes may interact in many different ways within an individual or interpersonally. These two types of entrainment, self-entrainment and interpersonal entrainment, engage different sets of coupling factors. In particular, entrainment to and through music needs to be seen as a particular case of entrainment in social interaction (Will and Turow 2011, 12–13).

Rhythm can be understood on one level as a wave frequency in neurodynamics and on another level as heartbeats and breathing expanding beyond the neurophysical containment of an individual body to produce interactions between bodies in speech, sound and touch, and in response to films, artworks and events in the world. Although ordinary everyday experience unfolds in this way, the philosopher Henri Lefebvre is acutely aware of how the rhythms of the everyday are produced by the ever-increasing power of society's time measurements and structural rhythms synchronising all aspects of human life. Artworks and films can copy these rhythms or break away from them, offering moments of uncertainty, stray thoughts and unusual narratives, which may or may not result in the resetting of dominant perceptual or emotional paradigms.

Empirical experiments show that rhythmic external stimuli, visual or aural, entrain delta and alpha oscillations. Flashing lights or pulsing sounds can reset neural oscillatory phases and entrain certain neural frequencies to pick out and follow the rhythms of such external stimuli. These examples of entrainment represent different kinds of perceptual and cognitive activity directly influenced by external stimuli through the mechanism of rhythm, which emerges as an efficient way to focus selective attention on the environment. Meanwhile, attention has been shown to be periodic; it oscillates and is susceptible to external rhythms visually, aurally and in the other senses. Thus:

Brain operation is profoundly rhythmic. Oscillations of neural excitability shape sensory, motor, and cognitive processes. Intrinsic oscillations also entrain to external rhythms, allowing the brain to optimize the processing of predictable events such as speech. Moreover, selective attention to a particular rhythm in a complex environment entails entrainment of neural oscillations to its temporal structure (Calderone et al. 2014, 300).

2.2   Metastability across Brain, Body and Art: Pollock

There are many studies that provide specific and detailed empirical support for the idea that internal neural dynamics produce features of chaos in cooperation with features in the world. Freeman and Skarda's work on the olfactory bulb of the rabbit shows that receptors 'allow the nose to capture odorant molecules at very low concentrations but, because the airflow through the nose is turbulent, the inhaled air carries the molecules to a different set of receptors with each sniff' (Freeman and Skarda 1990, 66). The internal neurodynamics are metastable and open to external molecular metastability described best using principles of chaos theory:

Odor information was then shown to exist as a pattern of neural activity that could be discriminated whenever there was a change in the odor environment or after training. Furthermore the 'carrier wave' of this information was aperiodic. Further dissection of the experimental data led to the conclusion that the activity of the olfactory bulb is chaotic and switches to any desired perceptual state (or attractor) at any time (Korn and Faure 2003, 789).

Different action potentials, perceptual routines and sets of responses are held in readiness but not committed to until something sparks a phase transition or criticality that will lead to the 'global' phenomenal identification of the odour or the learning of a new odour (plasticity): 'chaos confers the system with a deterministic "I don't know" state within which new activity patterns can emerge' (Korn and Faure 2003, 821). Rather than manipulating symbols or indexes of the world and its features, the brain and nervous system are continuously in a state of activity, in real-time contact with the surrounding world. For Freeman, the rabbit's olfactory system provides evidence that the brain's metastable activity allows it to be open to contingency. An external perturbation can be sampled and may trigger larger matching global patterns, routines and moments of stability or iteration (attractors). Thus, in the 'olfactory bulb, neither a neuron, nor a cell assembly, is designed to serve a single purpose. Rather, they can implement several functions depending upon the internal states of the brain and the constraints placed upon it by environmental factors' (828).

The important point is that chaos in the environment, the motion of particles and other contingencies comingle with the organism's biology, place, orientation, health, age and neuronal assemblages, together forming an immediate dynamic complexity or criticality, until relative stability is achieved. Examples of this stability might include identification, memory retrieval, contemplation or contentment. Freeman and Skarda show that the brain accommodates in readiness numerous dynamic systems that engage with the world, clearly illustrating the importance of a dynamic approach to understanding perception in art:

Perception is an interactive process of destabilization and re-stabilization via self-organized dynamics. Thus, we come to view the brain as the location where a self-organized process of patterning takes place, a process that reaches back toward the stimuli giving them form at the same time as it creates their biological meaning for the organism ... chaotic activity enables the rapid state transitions essential for information processing (Freeman and Skarda 1990, 279).[8]

While it is true that external perturbations have an effect on internal dynamics, chaotic behaviour can be self-generated by neurons and can help to shift and modulate perceptions, which are not fixed copies of external stimuli but creative interpretations. This already complex and variable set of events is complicated even further by neurotransmitters, such as serotonin, adrenalin, dopamine and oxytocin, that operate at the same time and 'can switch the activity of the cell from one stable firing pattern to the other' (Korn and Faure 2003, 796).

Another way of describing a stable pattern in apparent chaos is fractality. Fractal patterns (patterns that appear similar across different scales) have been observed in a number of modalities: thought and problem-solving, and language and eye movement behaviour and rhythms.[9] Isolating fractal patterns is a generalising or averaging process that picks out essential constants that repeat and which, at first glance, are not apparent. This implies that isolating fractals, or intuiting them, is a particular method of observation or attunement that can be studied or cultivated. Redies (2008) suggests various isomorphisms where artworks mirror specific patterns of encoding in the visual system. Weir and

---

[8] Importantly, Freeman and Skarda note:

[A] self-organized chaotic generator responds to environmental input by replacing it with an internally generated chaotic activity pattern. These self-organized chaotic activity patterns are transmitted further into the brain and provide the basis for future selectivity by (1) causing changes that mediate motivation, reinforcement and learning, and (2) modifying receptor input by causing direct environmental manipulation by the organism or by changing receptor positioning with respect to the world. The brain determines which input it will admit and what spatiotemporal form the resulting neural activity will assume. We suggest, therefore, that chaos is essential for input selection, processing and the creation of information in the brain (Freeman and Skarda 1990, 281).

[9] Paulson suggests that examples of word skipping, fixation duration and saccade length 'suggest that eye movements are iterative and are analogous to the fractal property of self-similarity at different scales' (Paulson 2005, 350).

Mandes write that 'eye movements when scanning an image follow a fractal pattern in time, that is, the geometrical fractal property is also characteristic of where the viewer's eyes go naturally when scanning' (Weir and Mandes 2017, 22). Torre explains that 'thinking and problem solving are fractal in nature because the structured irregularity of their processes and interactions exhibit statistically self-similar structure on all scales' (Torre 1995, 194).

We are fortunate to have numerous film sequences of the artist in the act of painting, which reveal the specific rhythmic manner that characterised his working method across a number of scales. Pollock clearly intended to use his whole body in the process of painting, placing large sailing canvases on the floor of a spacious barn. As though it were a kind of staking-out process, he begins by flicking paint onto particular parts of the canvas, sometimes adding to or crossing the first mark and then taking a stride or step to the left or right, repeating the process. Eventually he begins to join up the marks, allowing the paint to drip from the stick with more delayed, scribbling gestures (as if going over a line in crayon drawing) or with sinuous gestures allowing the paint to drip into various forms of the figure eight. Sometimes the time delay of paint dripping is extended and the movement concentrated in one place, creating tightly woven loops or circles, or the artist shortens the time delay with quick flicks to rid the brush of paint, creating explosive splash marks. This variable set of applications is revisited by adding more lines, marks and drips of paint, sometimes of a different colour, with the artist often returning days later to add new layers of activity. Sometimes cigarette ash, sand, pebbles, glass and other materials are added. Different kinds of paint, industrial and traditional oil paint, or gouache, are used. Watching footage of Pollock's process reveals a ceaseless and variable movement, few pauses, a continual poking action of the brush or stick into the can of paint held in the hand, bending over the canvas or kneeling, ferrying or helping to suspend the trajectory of paint from the can in a continuous movement, leaving languorous patterns on the canvas. The continuity of the body's motion with the paint and its motion seems important, as do the changes of direction and brief interruptions in smooth rhythms caused by the jerky action of flicking the paint from the stick or brush, which produces focal points that attract and slow down vision. This combination of actions resembles a Hollywood version of fencing, with parries and thrusts, jabs intercalated with sinuous moments, while sometimes it appears as if the artist is pouring sauce over food as a chef might do, involving the even-handed release of materials. In short, the painting is a rhythmic, itera-tive process across many levels of description.

Pollock's abstract expressionism links the brain, itself a metastability of memories, motives and improvisation, to the sensations and muscle memory of the body in an intuitive alertness to matter's movements in the world. This abstraction can to some extent be retrieved by the viewer, and certainly watching footage of the artist painting sensitises us to the vocabulary of rhythms, textures and gestures with which he worked. Along with actions that are taken in direct response to an artwork, film or piece of music, from rhythmic and arrhythmic eye movements to movements of the head, fingers, limbs and breath, such artworks can elicit action potentials: getting ready to act, respond or anticipate. These may not be acted out, but their readiness may be sustained in metastable ways:

[T]he elicitation of an action-readiness-pattern triggers a cascade of spatio-temporal dynamics in the brain modulated by sensory input that aids antici-pation on the interactions with the environment. In ballroom dancing for instance, the first measures of music will afford either dancing tango or waltz. The elicitation of the tango-dancing-pattern will trigger an attractor-manifold that governs the sensorimotor coordination between me, my dance partner and the music: this action-readiness pattern will make certain action possibil-ities solicit more to me than others. On a more fine-grained level, small cues by the dance partner and subtle variations in the rhythm in the music further specify my action-readiness. Only if I am well attuned to the context (the situation) and thus metastably poised for several relevant activities I could do next, can small cues in the environment lead to very different positions in state space and hence to flexible responsiveness to (very) different solicitations. That is, only when I am able to rapidly accommodate the small deviations from my anticipations ... can I engage skillfully with a complex environment (Bruineberg and Rietveld 2014, 8–9).

This explanation of action in the world as an externalisation of psycho-logical routines gives us an insight into Pollock's working method. And these psychological routines can be adopted by the viewer as they follow the flow of matter, gesture and rhythm in the artwork. We can go further with psychological description to see Pollock's working method as meta-stability that enables the individual to effortlessly switch between different patterns. Such metastability is essential in 'the brain-body-environment dynamics of skilled agents' (Bruineberg and Rietveld 2014, 8).

Bruineberg and Rietveld explain that action patterns can become clumped together as sequences, but there can also be improvisation in response to situations. For example, boxers have groups of punch combin-ations (or patterns) worked out, but they can also improvise between them. Similarly, musicians and artists have groups of action patterns that can be switched and adapted to specific requirements. Such actants do not have a fixed response, but a dynamic and continually changing openness or readiness. This is a 'metastable attunement' that allows any number of action patterns to be triggered or combined. To not have a pre-planned fixed strategy allows for a particular kind of improvisation and openness to chance. This can be called a 'context-sensitive selective openness' (Bruineberg and Rietveld 2014, 3), where the rhythms of materials and contingencies in the environment take their course with minimum interference from willed action. The body rhythmically alternates action routines that can self-organise into novel combinations, which, in art, can help to produce new images, techniques and paradigm shifts.

Paalen writes that Pollock's rituals of painting combine

an active part: assimilation through a sort of somatic mimetism (dance movements similar to the movements of animals, all kinds of mimicry of emotions), with a passive part: mimetism of camouflage (masks, disguises). These two kinds of mimetism, active and passive, are the two poles which release the great current of rhythm which, going as far as trance, in traversing the individual, effaces his personal memory in order to conjoin it emotionally with the great reservoir of generic memory (quoted in Belgrad 1998, 67).

Much of our fascination with Pollock's action painting amounts to a mental swaying with his body movements. Inklings of his whereabouts are suggested by the trajectories of paint, and there is a tacit engagement with patterns of energy, motor imagery and kinaesthetic involvement, which alternates with 'the image' (what the morass of lines, shapes, superimposed colours and densities bring to mind in terms of associations).

The art historian John Golding (2000) brings together a number of images from Inuit and Kwakiutl masks, and Mayan and Egyptian reliefs, to suggest possible sources for Pollock's earlier more figurative works, such as *Guardians of the Secret* (1943) (Figure 2.1) and *Man and Woman in Search of a Symbol* (1943).[10] It is tempting to think that Pollock's use of motifs, symbols and colours in this period gave way to a process of further abstraction, discarding any resemblance to these sources in favour of a distillation of rhythmic relationships, with sweeping semicircular movements that curl up into a spiral or areas punctuated by staccato dots and points. It is likely that Pollock sometimes worked, consciously or unconsciously, with mental images of his earlier paintings or that these mythopoetic sources were abstracted into rhythm and relationships of patterns passed into muscle memory, becoming routines of action and gesture.

[10] Stephen Polcari gives an account of Frazer, Campbell, Nietzsche, Jung and others as important thinkers who provided American Abstract Expressionist painters with 'individual statements of historically and culturally bound and formed metaphors or symbolic generalizations that constitute an open network of interlocking ideas, influences and events' (Polcari 1991, xxi).

Figure 2.1 Jackson Pollock, *Guardians of the Secret* (1943). Oil on canvas, 123.19 × 191.77 cm. Digital Image © Museum of Modern Art (SFMOMA), CA, USA. De Agostini Picture Library/Bridgeman Images © Pollock-Krasner Foundation/ARS. Copyright Agency, 2020. (A black and white version of this figure will appear in some formats. For the colour version, please refer to the plate section.)

As Golding writes, 'physical gesture and the pictorial symbol become synonymous' (Golding 2000, 126). Muscle memory can retain gestalts, rhythms and relationships abstracted from, but also re-enacting, these ancient forms. In fact, in the Jungian context, with which Pollock was familiar, the physical gestures may not be the result of the retrieval of a pictorial or ancient motif held in the conscious mind but the unconscious enactment of a 'rhythmic outline' – a motor image or action pattern passed into second nature, but which nevertheless resembles an ancient motif or symbol. This is what happens when we learn to autograph, writing our signature without consciously having to construct the letters and shapes because it has become automatic.

Pollock's most famous works are the result of tiny voluntary movements combined with larger ones, and walking around the canvas, but sometimes the paint seems to flow through his (passive) hands, through the suppression of the hands' movements. As we see in *Number 14 (Gray and Red)* (1948), the play of controlled dexterity alternates with being relaxed, disclosing the non-organised aspects of matter in particular ways. The viewer is able to alternate between these two poles while engaging with the work. The voluntary and involuntary are interwoven in visual ways, and for artists in tangible ways as well, since they actually handle the paint. In creating intricate patterns of linear complexity, the very fine bundles of lines only inches across where the artist allows the paint to flow in one chosen area, Pollock himself was an observer, watching the pattern of paint emerge from the end of the stick, seeing flickers of light reflecting off the glossy paint spiralling onto the canvas, perhaps with a childlike fascination for the way in which the paint seems to write itself and to self-organise into ever-tighter circles or looser rhythms. A studied patience and resolve is required to not move, to hold the hand over the canvas and let the paint collect there, becoming more complex and interwoven. A certain reverie may have accompanied this act, which viewers imagine they can share as they focus on a very localised area. It is not difficult to imagine how a particular pattern of a knotted bunch of lines could be kept in the mind as an after-image, complementing direct perception expanded over a larger scale in the same painting. That Pollock's rhythmic loops work on multiple scales, from tiny bundles to broad patterns, complements the fractal approach and suggests that the scales from the neural to the embodied and phenomenal are traversed by scale-invariant rhythmic outlines or schema, somewhere between stability and instability. In neuroscience, Ichiro Tsuda makes three points in this regard:

(1) chaotic activity at one level results from chaotic activity existing at a lower level; (2) chaotic activity at one level is independent of that at the lower levels, and rather it results from damped oscillations enhanced by feedback from activity at higher levels; (3) chaotic activity at one level results from a self-organization at the lower level (Tsuda 2001, 837).

Tsuda maintains that chaotic activity at the phenomenal level results from chaotic activity existing at the lower neurological level but occurs in a reciprocal manner so that the higher level also effects a change in lower-level dynamics.[11] 'Chaotic activity' is not stochastic and should be understood as complex dynamics with novel elements as well as elements of recurrence. The term 'chaotic itinerancy' describes this mixture of stability (memory, forms) and instability (formlessness, novel associations, perceptions not held in memory). Chaotic itinerancy is analogous to

the cyclical visits of itinerant workers to familiar places with seasons and years. Each new site of visitation is governed by an attractor that dissolves into 'ruins' (after-effects) even as soon as it is actualized, persisting in its pervasive influence

---

[11] For an overview of how memories may be organised along these lines, see Tsuda (2001).

but allowing the brain state to avoid capture and incarceration in some patho-logical deep well (Freeman 2001, 816).

The cyclicality observed in neural populations and their oscillations (par-ticularly in memory areas of the hippocampus) and the way in which memories appear to re-occur during the perceptual process of looking at a painting show that memories themselves, although relatively stable, are not stored as fixed codes but are dynamic 'routines' which can entrain other dynamics at different levels. It is important to remember that chaotic itinerancy can also be understood in terms of rhythm and iter-ation, or as a complexity of rhythms, the pattern of which can be repeated at different levels of description.[12]

Tsuda published a 'rhythmic diagram' of chaotic itinerancy, a kind of map of a thought process where a memory is retrieved while there is ongoing perception taking place (he does not specify what is being per-ceived) (Figure 2.2). Within the perceptual space there may be prior events or memories that are triggered along with semantic associations held in the memory. The artist Lee Krasner writes: 'All my work keeps going like a pendulum; it seems to swing back to something I was involved with earlier, or it moves between horizontality and verticality, circularity, or a composite of them. For me, I suppose that change is the only constant' (Krasner in Tucker 1973, 8). The dynamic and complex neural substrate for this process can be described in terms of chaotic dynamical systems. Tsuda refers to fragmentary memories as 'attractor ruins'. In viewing a painting by Pollock, for example, these attractor ruins could be understood as traces of the artist's iterative movement or gesture, to which we keep returning: Pollock is a 'prior' presence that resonates with the ongoing dynamics of our perceptual process. There may also be iteration patterns, a combination of colours or a rhythmic outline, that

---

[12] This is somewhat different from Jones's argument that Pollock's 'repetitive' machine-like painting style emerged as a practice in keeping with the historically specific conditions of capitalist industrial production and Taylorisation, equally influential for, and amenable to, Greenberg's evaluation of abstraction using rational positivist methods. The more recent discussion of the fractal aspect of Pollock's work may help to explain how the repetition, so characteristic of the age of capitalism, was internalised somatically in muscle memory and practice from the external spatio-temporal environment – perhaps even unbeknownst to the artist. But it seems important to point out that there are two poles to this internalisation: one is the order and repetition of the historical condition (amenable to rationalist art criticism); the other is spontaneity and innovation, which emerge within these constraints and which find their correspondence with Rosenberg's notion of action painting as a kind of existential struggle. The pole of order is rhythmic repetition, the pole of 'disorder' is rhythmic complexity – jazz. The relationship between the two poles produces a metastability in many artists' practices. It also allows the two different kinds of art criticism (one stressing cognitive order and the other chaotic contingency) to be understood as descriptions of the same phenomenon.

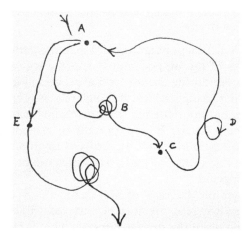

Figure 2.2 Author's drawing of diagram of chaotic itineracy after Ichiro Tsuda (2001). 'Toward an Interpretation of Dynamic Neural Activity in Terms of Chaotic Dynamical Systems'. *Behavioral and Brain Sciences* 24 (5): 799.

seem to appear in other parts of the canvas.[13] Recognising this similarity while perceiving differences requires both stability and metastability – and an openness to differences. For Tsuda, attractors are compelling perceptions. In some of Pollock's paintings, the artist repeats routines or patterns such as figure eights, lozenges and interstices filled in with colour. 'Itinerancy' or nonlinear activity suggests the interaction between attractor ruins and attractors: Tsuda is interested in giving an account not of the states themselves (a memory or a perception, a lozenge) but of the numerous dynamic relations among them. In other words, 'Tsuda appears to adopt a variant of the Bartlettian framework, noting that memory is the product of a complex interplay between what is stored and what is currently being perceived' (Foster 2001, 816). This inter-weaving is represented as the joining of a relatively stable order (attractor ruins or memories) and ongoing possibilities of perception, with an openness to external perturbations that are yet to become stable.

---

[13] For an analogy in language, see Hasson et al. (2015), who suggest that aspects of implicit memory are co-present with unfolding speech comprehension. Implicit memory has been associated with DMN activity (Yang et al. 2010, 354). Hasson et al.'s research into the co-presence and anticipation involved in communication shares important synergies with Tsuda's attempt to conceptualise the co-presence of attractor ruins and attractors. Hasson et al.'s work on film has some potential for Tsuda's research to be applied to the visual domain, which, I am suggesting here, would benefit greatly from this treatment.

It is difficult to decide if the constants or the aberrations provide more of an aesthetic effect, the forces of composition or diffusion. It may be that it is the dynamic itinerary in between, that metastability that is produced, that creates the aesthetic play. Is metastability an aesthetic category? It probably should be, but only at the expense of undoing traditional aesthetics. I believe that metastability is fundamental in our experience and understanding of abstract art. This does not mean vanishing into the ineffable or into the uncountable. Rather, it means we can switch between known points and value the wandering routes taken in between.

Tsuda's chaotic itinerancy gives us a framework for understanding Pollock's holistic and irreducible process: a fluid, nonlinear dynamism between points. This complements studies of fractals in Jackson Pollock's paintings (Taylor et al. 2011). A brief discussion of Taylor et al.'s work on Pollock is a good way to consider how internal and external chaotic dynamics are sustained across brain, body and world as system coupling, rather than being all in the head. According to Taylor et al., viewers might be attracted to fractals because they are relaxing. Alpha-band activity accompanies the viewing of fractals, which he suggests demonstrates an instinctive appreciation for nature, which also has fractals. But perhaps the later stages of experiencing the artwork can be more accurately described as putting the alpha-band activity, seen in mind wandering, into contrast with the more evaluative frame of mind that sometimes intervenes. This would then produce a dynamic switching between relaxed and analytical states of mind and different kinds of vision: looking hard at local, fine-grained details with consecutive vision (gamma), and relaxing focus and diffusing details for coarse- or indefinitely grained configural vision (alpha). Perhaps this would create a map of chaotic itinerancy involving phenomenal (aesthetic) as well as neurological patterns. According to Van Leeuwen (1998), chaotic dynamics supports the rhythmic flipping of shapes in perceptual multistability. But Cupchik surmises that switching from detecting edges to observing the painterly colours of planes and blobs (and, by extension, switching focus) is an aspect of fractality, as we see with Pollock's abstract painting, which can be 'soothing' (Cupchik 2016, 293), a term that suggests the more basic principles of rhythm and the oneiric.[14]

---

[14] In viewing one of the best-known illusory diagrams, the reversible transparent cube (the Necker cube), viewers 'experience a spontaneously changing percept, alternatively seeing either the bottom or the top of the cube. Since the input remains constant, it is our brain that does the switching by some rules' (Buzsáki 2006, 275). Viewers who pressed buttons every time they experienced a switch felt that they were doing so in a random manner but the results suggested a memory effect where viewers were rhythmically switching in ways that suggested an underlying stability.

It is interesting that the perfectly mathematical fractals we see represented in diagrams are not as attractive or as soothing as fractals that are 'messier'.[15] It is as if when the underlying logic of a pattern is discovered it becomes less interesting, and perhaps even menacing, as we see with rumination and feverous visions where repeat structures and aporias rise up in the cracks on the plaster. These messier fractals can be experienced as chaos-becoming-order or order-becoming-chaos.

Clyfford Still, whose huge effulgent paintings have been compared to images of ancient plumed serpents or tongues of fire, adopted the use of self-same forms in *January 1947* (1947) (Figure 2.3). Here the artist repeats hooked forms on a number of different scales and in different colours, capturing a flickering, shimmering quality that is both light and matter, 'an oddly crusted luminosity ... [with] jagged slabs of pigment'

Figure 2.3 Clyfford Still, *1949* (1949). Gouache on paper. Digital Image © Private Collection/Bridgeman Images © Clyfford Still/ARS. Copyright Agency, 2020.

---

[15] See also Purcell et al. (2001), which suggests that subjects' preferences for images are associated with fractal geometry.

(Golding 2000, 126) that are built up with repeated gestures of the hand and wrist across the canvas. It is to the artist's credit that these forms can be read as torn fragments of paper or parchment, flames, plumes or jagged rock, surges of sensation, or licks of paint – different scales and textures within which a repeating pattern is nested. In case this account of Still's work edges too close to the coldly technical, we have recourse to Bachelard. He writes of a certain 'calorism' that 'corresponds to the materialization of a soul or to the animation of matter; it is a transitional form between matter and life' (Bachelard [1938] 1987, 75). Any attempt to reduce Still's paintings to the abstraction of motifs based on feathers and fire, as seen in the art of shamanistic ritual, the Chinese flaming pearl or the Persian simurgh, would rationalise abstraction by giving priority to the decoding of semiotics and the study of myth. Bachelard rejects James Frazer's explanation in *The Golden Bough* (1890) of the Hawaiian origin of fire, writing that it

serves to explain the particular colour of a certain species of bird . . . such explanations, hypnotized by an objective detail, fail to take into account the primitivity of the affective interest. The primitive phenomenology is a phenomenology of affectivity: it fabricates objective beings out of phantoms that are projected by reverie, it creates images out of desires, material experiences out of somatic experiences, and fire out of love (Bachelard [1938] 1987, 37–38).

Rather than providing rational-functionalist explanations for the fascination afforded by observing fire, Bachelard associates this fascination with reverie and desire, which gain added momentum from the oneiric and provide rhythm to the matter in art. Bachelard reveals what is possibly at stake in looking too deeply into the fire, its purity and danger:

Since we must disappear, since the instinct for death will impose itself one day on the most exuberant life, let us disappear and die completely. Let us destroy the fire of our life by a superfire, by a super-human superfire without flame or ashes, which will bring extinction to the very heart of the being. When the fire devours itself, when the power turns against itself, it seems as if the whole being is made complete at the instant of its final ruin and that the intensity of the destruction is the supreme proof, the clearest proof, of its existence. This contradiction, at the very root of the intuition of being, favors endless transformations of value (Bachelard [1938] 1987, 79).

It is not incidental that a flickering fire has fractal properties and the reverie that attends our viewing of the fire seems to be attuned to its rhythmic iterations. This attunement could explain the metastable substructure of the aesthetic experience we might have in viewing many of Still's vast canvases.

There is something flame-like in Pollock's *Stenographic Figure* (1942), a writhing beast engulfed in flickerings of stenographic code that promise

Figure 2.4 Jackson Pollock, *Stenographic Figure* (1942). Oil on linen, 101.6 × 142.2 cm. Mr. and Mrs. Walter Bareiss Fund. 428.1980/ Museum of Modern Art, New York. Digital Image © Scala, Florence © Pollock-Krasner Foundation/ARS. Copyright Agency, 2020. (A black and white version of this figure will appear in some formats. For the colour version, please refer to the plate section.)

communication, yet we continually return to the rhythms of the image. The title, referring to stenography, is already a cultural form in our semantic memory and becomes a relatively stable interpretative frame held in the working memory while visual inspection takes place. Stenography is writing in shorthand, a code used for transcribing speech (Figure 2.4). Since only a few symbols in shorthand can be used instead of several letters of a word or phrase in longhand, it allows the stenographer to keep up with speech so as to be able to transcribe it later in full. As such, stenography can be understood as a code that compresses or parses speech and, by extension, compresses or parses action. This is already a complex way to pique the interest of the viewer who might understand abstract art as a kind of code. But the title also refers to a figure, and the viewer is primed to search for it amid the stenographic marks. This triggers neural action potentials and facial recognition structures. Top left, on a black ground, an animal's head can be discerned with its jaw wide open, and in other parts of the painting there are numerous fingers or claws, associated with rapid finger movements and the artist's own dexterity in creating rhythmic squiggles and gestures. We read a quickness in the application of paint, the larger swirling gestures associated with cursive script within which are more fine-grained manipulations of the hand, all apparently trying to keep up with speech or thought which the 'writing' is supposed to transcribe.

The artist creates an opportunity to think about the tension between 'semiotic' impulses where the brain, primed to understand visual experience as a set of symbols representing prior meaning, sparks linear reading routines from left to right, versus the nonlinear 'asemiotic' aspects of the work, the affective, rhythmic, flame-like qualities of sensation, colour and gesture that are accessible through the senses before they are codified into language.[16] An 'inner beast' surfaces, the head of some animal against the black background claws at the twisted body and buttocks on the table, and everything seems to writhe in flames. The semiotic world of stenography tries to explicate, or keep up with, the primordial energy of creative expression, while at the same time being engulfed in its passion.

All this tangled thought and sensation we experience within the perception space afforded by the painting is a nonlinear dynamic process, not too controlled, nor too chaotic, and in various patterns of interaction. Although I have explained in writing, with linear progression, a possible way to experience this painting, while observing the painting, my thoughts and sensations do not occur in a sequential manner but rather as a back-and-forth movement between stable points (attractors) and departures from these points (attractor ruins), such that a chaotic itinerancy takes place. Tsuda's 'map' is a simplification (a codification) of the complex process of iteration, mind wandering and innovation that occurs in the perceptual space extended by Pollock's painting. The fact that Tsuda's map resembles an automatist drawing by Masson or the lines in Pollock's paintings is significant: both the scientific diagram and the artworks help us to think beyond the rational and sequential ordering of events. An apposite and intuitive understanding of this automatism can be found in Katherine Conley's description of the 'rhythm of automatism characterized by alternations between moments of suspension and moments of flow' (Conley 2013, 8). The metastability of mental phenomena mobilised in producing an automatist drawing may have attractors, attractor ruins and chaotic itinerancy in between. The sinuous line in

---

[16] Archaeologist Carl Knappett writes:

It is as if us humans interact with objects using two enormously different logics simultaneously, the one linguistic, codified and symbolic, the other embodied, uncodified and pragmatic. Yet such a deep cognitive division seems highly improbable, given the often-seamless interaction between the human mind and the material world (Knappett 2005, 49).

autonomism not only expresses this metastability but also helps to produce it in observers as they follow, with their own rhythmic saccades, the meandering line pausing or returning to 'ruminating' knots. Something of this kind happens with Tsuda's diagram as well, a remarkable confluence of science and art. Pollock's working method of laying down a base layer of lines and knots, often in black, before painting over lines in other colours also points to this procedure (Figure 2.5).

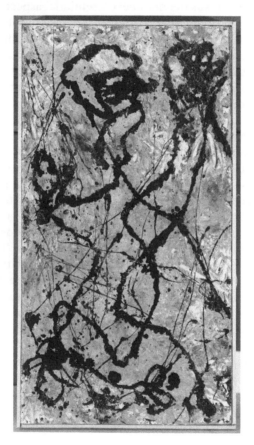

Figure 2.5 Jackson Pollock, *Untitled (Composition with Black Pouring)* (1947). Oil and enamel on canvas mounted on masonite, 43.8 × 23.3 cm. Oil and enamel on canvas mounted on masonite, 43.8 x 23.3 cm. The Olnick Spanu Collection. Digital image © 2020 Christie's Images, London/Scala, Florence © Pollock-Krasner Foundation/ARS. Copyright Agency, 2020.

Tsuda's neurodynamic approach alerts us to these multiple events and processes and sharpens interpretative analysis. It also provides considerably more detail about the general concept of 'absorption' in art proposed by Michael Fried. This absorption is extremely heterogeneous and iterative, broken by moments of the opposite tendency (detachment). It joins relatively stable recurrent dynamics (prior patterns of memory and percept formation) with ongoing perceptual contingencies. These contingencies are tracked best by neural metastability and passive observation. It is interesting that Fried's absorption and Heidegger's *Dasein* are both related to the immersive properties of reverie. In Fried's later work, this is explicitly acknowledged.[17] What is significant is that art historians, philosophers and psychologists seem to be describing the same phenomenon. The psychologist Mihaly Csikszentmihalyi concurs: '[P]hilosophers describing aesthetic experience and psychologists describing flow are talking about essentially the same thing' (Csikszentmihalyi 1990, 8). Csikszentmihalyi's well-known studies of the flow experience help to explain how the encounter with art has multiple dynamic features which together can be understood as a form of immersion. His model contains the following salient features: (1) attention centred on activity, (2) no awareness of past and future, (3) loss of self-consciousness and transcendence of ego boundaries, (4) skills adequate to overcome challenges, (5) clear goals and feedback and (6) 'autotelic' nature (does not need external rewards, intrinsically satisfying). While features (3), (4) and (6) are comparable with the reverie involved in abstract art, and matter painting in particular, feature (5) seems to be at odds with (6). As with many aesthetic models of this nature, Csikszentmihalyi emphasises heightened consciousness related to mindful involvement that is directed towards unity and harmony. This is somewhat different from the nonlinear, involuntary mind-wandering aspects of drift and arbitrary contingencies, enjoying chaotic ripples that can be understood as alive and exciting. This drifting is 'autotelic' in nature (6).

Tsuda's diagram is important because it gives us a glimpse of what chaos in the brain might 'look like' and shows us what Csikszentmihalyi's flow state might entail if properly attuned to the nonlinear dynamics of experiencing art. Tsuda's diagram illustrates in a simplified manner how nonlinear thought is isomorphic with doodling and automatist flows, which lead to unexpected destinations, complicated by recurrent and itinerant resting points. This establishes the importance of nonlinear dynamics as characteristic of the thought process involved in automatism and forms of abstract expressionism, in the viewer's sensitivity to these

---

[17] See Fried (2008).

practices and in the art historian's account of this relationship. This is considerably complicated by Figure 2.4, Pollock's *Stenographic Figure*, but in ways that are much more than just a metaphor: while Pollock's painting is the product of nonlinear rhythms, starts and stops, we can pick up and elaborate these rhythms with our own creative itinerancy. This form of visualising the rhythms and iterations of thought in both science and art provides an insight into what might be involved in Ehrenzweig's notion that, in viewing art, 'outward perception and inner phantasy become indistinguishable' (Ehrenzweig [1967] 1993, 272). This is because the artwork provides a set of visual marks, with intervals, colour variations and perceptual ambiguities that are felt as dynamic *and* somatic; they influence the rhythms of eye behaviour and thoughts that arise nonlinearly. This is an important way to understand how an ancient motif can become a rhythmic outline, re-performed through muscle memory in the act of painting. We can imagine that these rhythmic outlines or routines, sometimes appearing fractal, link muscle memory to cultural memory in the execution of the outline. Attractor ruins, attractors and chaotic itinerancy may be seen as a way to describe these patterns underlying different perceptions and sensations. This is complicated further by the fractal flickering of the fire which produces reverie. According to Bachelard, the hypnotic state into which fire is able to send us seems surprisingly immersive, emphasising the fact that external elements can entrain non-intentional thought and mind wandering, a kind of dreaming with eyes open. This kind of reverie

is entirely different from the dream by the fact that it is more or less centred upon one object. The dream proceeds on its way in linear fashion, forgetting its original path as it hastens along. The reverie works in a star pattern. It returns to its centre to shoot out new beams. ... It is a phenomenon both monotonous and brilliant, a really total phenomenon: it speaks and soars, and it sings (Bachelard [1938] 1987, 14).

There is something of this magical and creative phenomenon in the work of Clyfford Still (Figure 2.3).

Ehrenzweig associated creative thought with 'radiating pathways' or bifurcations, which he published as a diagram. He explains it in the following manner:

A creative search resembles a maze with many nodal points. From each of these points many possible pathways radiate in all directions leading to further crossroads .... Each choice is equally crucial for further progress. The choice would be easy if we could command an aerial view of the entire network of nodal points and radiating pathways still lying ahead. This is never the case. If we could map out the entire way ahead, no further search would be needed. As it is, the creative thinker has to make a decision about his route without having the full information needed for his choice. The dilemma belongs to the essence of creativity (Ehrenzweig [1967] 1993, 37).

On this view, mind wandering at the macro level of creative thought shares metastability and patterns of 'wandering' with lower levels of chaotic itinerancy. Smithson writes: 'There's a sort of rhythm between containment and scattering. An artist in a sense does not differentiate experience into objects. Everything is a field or maze, and you get that maze serially.... The seriality bifurcates; some paths go somewhere; some don't' (Smithson 1979, 168–169). Automatist drawings, Ehrenzweig's diagram (Figure 2.6) and Bachelard's observations on reverie all derive from the expression of a nonlinear stream of consciousness. This kind of rhythmic outline is important for creative thinking, unaffected by linear rational exposition. The Tsuda diagram, extrapolated from computational experiments based on the behaviour of neural populations, is also a visualisation of nonlinear dynamic thought. On the level of philosophical discourse, this kind of

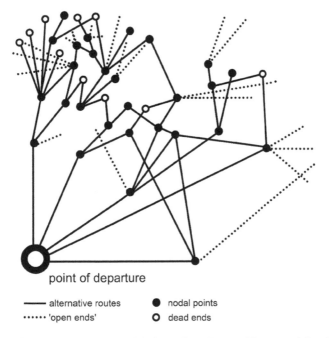

point of departure

——— alternative routes   ● nodal points
•••••• 'open ends'   ○ dead ends

Figure 2.6 Author's drawing after Anton Ehrenzweig's diagram of a creative search process, in *The Hidden Order of Art: A Study in the Psychology of Artistic Imagination*. London: Weidenfeld and Nicolson (1967), 36.

nonlinear dynamic complexity has been recognised by Deleuze in his use of the term 'mapping' as a way to think about creativity, one of the most important Deleuzian themes. For Deleuze, a 'map' of creative thought does not simply trace a set of given or premeditated coordinates that are followed, but is a way of opening up imaginative routes, a psychogeography that improvises with the features of the painting, building in instability and spontaneity in order to make new connections rather than tracing old routes.

This view of mapping as a creative and rhythmic itinerancy, which is very close to Ehrenzweig's understanding of creative practice, finds equivalents in the psychogeography of the Situationists and the philosophical walking of Heidegger's *Country Path Conversations* (2010). The somatic dimensions of Pollock's and Masson's automatism are also a form of mapping in Deleuze's sense of the term.[18] At the level of experiencing sensations (scattering sand, for example), metastability releases matter running through the hands, or it takes the line for a walk – it is the brain's own metastability, externalised in the world ('acting in' the world) through the body's motion. As Ehrenzweig writes, 'the creative thinker has to advance on a broad front keeping open many options ... without being able to focus on any single possibility' (Ehrenzweig [1967] 1993, 36). A map might record some of these peregrinations and experiments, following routes and dwelling points, but, as these different levels of description indicate, it is best used to explore future possibilities. This rethinking of cartography helps us to understand that artworks can also act as 'maps' which record various cognitive and somatic patterns and offer opportunities for further elaboration by the viewer.

Studies of metastability in the brain emphasise that the brain is in constant flux, 'its dynamic ensembles ever rearranging themselves as processes unfold that weave immediate and past events at numerous temporal and spatial scales' (Tognoli and Kelso 2014, 36). As noted earlier, Tognoli and Kelso suggest an 'external perturbation ... can shift the coordination dynamics' (36) poised 'at the border between order and disorder' (38). I find this particularly compelling for understanding the occurrence of an extended period of relatively stable thoughts and associations, which are drawn from cultural and biographical memory:

---

[18] When Deleuze refers to *Informel* as somewhat 'like a map as large as a country' (Deleuze 2003, 104), referring to Jorge Luis Borges's short story where such a map of the world is as large as the world it features, the three-dimensional, analogical, sensory and phenomenal world becomes 'its own best model' over and above any parsing or codification of it. Compare this concept with Smithson's statement about his *Site/Non-Site* works, heaps of earth and sand in the gallery, that he regarded as a three-dimensional map.

a painting by Pollock can be seen as an autumnal landscape with twigs, roots and branches or as a galaxy full of stars and comets.[19] Yet the coordination dynamics can shift to more chaotic thoughts about the matter, viscosity and gloss of the congealed paint, which can lead deeper into analytical thought about technical aspects of the artwork or to another spell of mind wandering based on the appearance of textures and colours.[20] The shift from mind wandering (delta frequencies) to attentive observation (gamma) in itself suggests an interesting metastability. As Ehrenzweig writes:

It is extremely difficult to hold onto the interludes of dream-like ambiguity and broader focusing that are interspersed among the sharper images of conscious memories. It requires perhaps singular powers of introspecting into stream of consciousness to remember or rather reconstruct the innumerable twilight states .... It is possible to train one's powers of introspection to hold onto the less articulated states of consciousness and to earlier phases in the history of perception (Ehrenzweig [1967] 1993, 87–88).

2.3   The Aesthetics of Mind Wandering

Traditional approaches to mind wandering assume it is a deficiency, an obstacle to learning in classrooms (Smallwood et al. 2007), where working memory resources are diverted to the processing of internal representations rather than attending to external details.[21] However, this may be true only for certain kinds of learning tasks. Reading a detective novel, solving a puzzle and reading a map are demanding tasks that require the application of learned procedures and rational intentions, often followed in progressive or linear fashion. An artwork or poem involves different learning processes and outcomes. Whereas reading a novel or solving a puzzle can be purposeful and pleasurable, these activities require that we minimise distractions and keep an eye on possible interpretations, evaluating each as they occur to us. Here mind wandering needs to be minimised in order to reduce the load on working memory engaged with relevancy checks for the problem at hand. Relevancy checks require matching present perception to prior requirements. Imagining situations

---

[19] Clark (1999) suggests the metaphor of stars and comets for Pollock. This is counter to Greenberg's artificial constraint that viewing modern art should be about the contemplation of abstract, *a priori* forms.

[20] It is interesting how these two different kinds of 'natural' psychological responses find ideological ramifications in art criticism. For some interesting controversies, particularly those between formalist and social history approaches to abstraction, see Varnedoe and Karmel (1999) and Frascina (2000).

[21] The following opening paragraphs on mind wandering were first published in Minissale (2013, 240–244).

where relevancy might be postulated could involve aspects of mind wandering, but the kind of mind wandering I am interested in is closer to daydreaming and allows us to discover new things about art.

The default mode network (DMN) will be differently engaged depending on the kind of mind wandering involved, and there is much sliding between transitional states involving or recruiting aspects of non-DMN networks to create different experiences of mind wandering. One of the key things to note about the DMN is that it is not a unitary system but differentiates, that is, behaves differently, 'according to task and even type of stimulus' (Sreenivas et al. 2012, 229). This suggests that different examples of art stimulate variable structures of DMN activity. Studies by De Bruin and Kästner (2011) and Ellamil et al. (2012) show how different networks can work together to produce fluctuating states of mind wandering. Ellamil et al. (2012) show that mind wandering sometimes recruits brain regions such as executive networks that are associated with focused evaluative thought, held to be the opposite of what we understand of mind wandering.[22] But rather than viewing mind wandering as a failure in attention, it should be acknowledged as a form of aesthetic experience of abstract art. As Fazelpour and Thompson state, 'the purely negative, operational characterization of spontaneous, self-generated thought as "task-unrelated" fails to distinguish among the variety of such forms of thought' (Fazelpour and Thompson 2015, 225). They identify

the need for both more phenomenologically nuanced conceptualizations of the various forms that spontaneous cognition can take, and for methods that can provide tighter couplings between the contents and temporal dynamics of spontaneous cognition and the activations of given brain regions and networks (225).[23]

Fox et al. write that mind wandering can be 'tinged with fantasy not restricted by a single topic which might demand focused, purposeful attention and planned outcomes. Such thought has been likened to

[22] The work presents a good overview of the literature on mind wandering, but one of its major contributions is to suggest that the idea of a 'mental state', which neuroscientists are wont to identify as part of the family of spontaneous thought to which mind wandering belongs, is misleading. A 'state' suggests something stable and distinct: mind wandering is dynamic and recruits different brain areas within and outside the default mode network, with many fluctuations in attention to external and internal contents. The fluctuations themselves are characteristic of spontaneous mind wandering over a period of time.

[23] Even our assumptions about the all-important phenomenon of 'attention' should be attenuated by an understanding not only of contingencies but of the thing that is actually being attended to. Mole (2011) argues that attention cannot be understood by identifying certain neurological processes associated with instances of attention because whether a subject is attentive is not driven by these processes but, rather, depends on whether the processes are matched by the agent's task. He suggests that attention can recruit a broad and heterogeneous range of resources, depending not on the nature of attention but on the nature of the task – and obviously there are an infinite number of tasks.

daydreaming and to meditation' (Fox et al. 2013, 1). Various psychological studies suggest that creative insights occur in the 'resting' state (default mode network) when the mind is decoupled from demanding tasks that require our full attention (Ellamil et al. 2012). Furthermore, De Bruin and Kästner note that

the agent may engage in 'offline' processing, while controlling for or suppressing 'online' processing, to achieve the best result. Focussing on direct coupling (online processing) alone neglects the agent's ability to generate new and more advantageous conditions for relating to her environment (De Bruin and Kästner 2011, 7).

Many of these studies suggest that different kinds of mind wandering can happen with varying degrees of self-reference, which may involve relatively disorganised thoughts and that there are personal memories supported by the parahippocampal cortex involved with other kinds of mind wandering. Certainly, personal memory has been shown to be active during artistic production (Kowatari et al. 2009). Alternatively, there are times where one is simply not conscious of mind wandering and is hard put to recall its intricacies.

The default mode network and executive (control) network have often been thought to be mutually exclusive or inhibiting, but it is clear that their operations can overlap. Mind wandering can often arise as a twilight zone between daydreaming and focused attentiveness (Ellamil et al. 2012; Dixon et al. 2014). Dada and Surrealist artists viewed mechanistic, rational and intellectual motivations for art as suspect and inferior compared to the power of affect and unexpected irrational leaps and juxtapositions. Yet even non-conscious thought in daydreaming and mind wandering depends on conceptual combinations and structuring relations, which also happens during dreaming. Dreams allow unusual 'irrational' narratives and structuring relations between concepts, perhaps because of the deactivation, for the most part, of areas of the prefrontal cortex (Muzur et al. 2002) that we recruit for relevancy checks.[24] These areas usually exert 'rational' control and assign priorities for working memory, helping to make sense of conflicting stimuli or to ignore irrelevant details that do not conform to a situation type or long-term goals. Importantly, these prefrontal areas are responsible for the temporal order or sequencing of the retrieval of situation types. If the timing is skewed, objects will appear in strange situation types or contexts. In mind wandering, the syntax of conceptual combination is spontaneous

---

[24] This may suggest clear functional separation between the prefrontal areas and the default mode network, when in fact recent fine-grained analyses reveal a complex set of inter-relationships between control and relaxing control in the mind (see Dixon et al. 2018).

and non-rational and, as the Surrealists hoped for, is sometimes poetic and imaginative, speaking to desires and fears usually blocked by rational control or purposeful tasks.

When the executive control network and 'mindfulness' of metacognition are not functioning optimally because of exhaustion through physical exercise, the influence of drugs or alcohol, or even through religious rituals of exhausting energy by repeating words or actions, 'surreal' affects and conceptual juxtapositions are common. This principle of 'irrational' categorisation is precisely what happens in Surrealist art and film. The uncanny or uneasy feeling we might have in experiencing such art (think of Meret Oppenheim's *Déjeuner sur l'herbe* [1936], the fur cup) must have something to do with how it induces 'irrational' dreamlike non-sequiturs, making us uneasy because, although we are fully aware and reality-checking, orientation by trying known situation types does not seem to help. Alternatively, we are happy to go along with nonsensical or poetic thoughts and metaphors, a kind of mind wandering which creates a positive mood. Such works are a reminder of the principles of irrational conceptual combination which we customarily adopt in dreaming thoughts or which occur with sleep deprivation. As with Leonardo da Vinci's surrendering to abstract shapes and forms that suggest objects or scenes, when one 'surrenders' to the shape-suggesting forms of abstract art without forcing upon them a strong metacognitive analysis, free association between ideas can emerge and it can be unpredictable. One would need an evaluative process at some point, involving some metacognition, in order to select images or associations for later use, which can be important in many artistic practices. But the ability to zone out, as an artistic practice, as a creative process of thinking, can be cultivated.

It seems that what these different kinds of activity have in common is that they dispel rational analysis, self-awareness and inhibition in order to allow randomness or chaos in the connections between thoughts. This habit can be cultivated, requiring patience and passivity as well as some degree of attentiveness. As Corballis suggests, the distinction between mind-wandering and mindfulness

is not absolute. One of the techniques of mindfulness or meditation is to focus attention on the body, starting with the feet and moving slowly upwards. This is indeed a constrained wander, although hardly a walk in the garden or a stroll along the beach. It may well be that mindfulness is a means of resting the wandering mind, energising its resources (Corballis 2014, 147).

It may also be that exhausting the executive functions more easily allows involuntary thought to emerge. This is what happens during spiritual practices such as chanting or whirling, where the voluntary conscious seriality of repetition jams the executive network.[25] This also happens in many performance artworks and can be seen in Pollock's working method. The artist Antoni Tàpies describes one particular experience in his art-making in terms of questions rather than certainties:

The most sensational surprise was the sudden discovery, one day, that my pictures, for the first time in history, had become walls. By means of what strange process had I arrived at such precise images? And why did they make me, their first viewer, quake with emotion? ... Was it the culmination of a process of fatigue brought about by the proliferation of an easy tachism in the world? A reaction to escape anarchic informalism? An attempt to flee abstract excess and the urge for something more concrete? (Tàpies 2013, 20)

Thus mindfulness becomes an acted-out repetitiveness, an obsessive-compulsive ritual, sustained voluntarily until one is able to enter a mind-wandering state where the imperative to repeat goes on autopilot and one can entertain mental images and thoughts without self-consciousness.

One sees this with the predominance of the grid form in modern art. The repetition is machine-like, rhythmic and ruminative, and we see it in the works of Andy Warhol, Agnes Martin, Sol LeWitt and many other artists. Krauss provocatively suggests that artists who use grids in their work are substance dualists. Piet Mondrian and Kazimir Malevich, for example, were not interested in 'discussing canvas or pigment or graphite or any other form of matter. They are talking about Being or Mind or Spirit. From their point of view, the grid is a staircase to the Universal, and they are not interested in what happens below in the Concrete' (Krauss 1979, 50). And later, more forcefully, she states: 'I do not think it is an exaggeration to say that behind every twentieth century grid there lies a trauma that must be repressed – a symbolist window parading in the guise of a treatise on optics' (59). Yet this analysis is itself a logical reductionist understanding of the grid in modern art, blocking off imaginative, affective and poetic associations which do not have to be understood as 'spiritual', but rather as serotonergic. Repeated forms stimulate pleasurable and rhythmic possibilities until they reach a surfeit and become monotonous – another trigger for mind wandering – at which point the mind drifts into involuntary or arbitrary associations whose comings and goings may be buffeted by desires or memories. The abstract grid can be transformed from a prison of thought into a playground of sorts, a visual support not for logical optics but for the free play of mental imagery.

With the work of Agnes Martin, we surrender to involuntary abstractions. In a view of *White Flower* (1963), the arrangement of matter in the

---

[25] It should be noted, however, that many religious traditions denigrate mind wandering and spontaneous thought as frivolity and moral weakness. See Eifring (2018).

Figure 2.7 Agnes Martin, *White Flower* (1960). Oil on canvas, 182.6 x 182.9 cm. Solomon R. Guggenheim Museum, New York, Gift of Lenore Tawney, 1963. Digital Image © Guggenheim/Scala, Florence © Agnes Martin/ARS. Copyright Agency, 2020.

visual field is repetitive, a series of white hand-painted rectangles on gold ground (Figure 2.7). The meticulous and repetitive process seems to be the product of mindfulness – focus and concentration, purpose and will. Yet while viewing these grids, supported by the resolve of the prefrontal cortex, the sense of rational will involved in counting tends to dissolve, just as in meditation the counting of one's breath subsides. In both cases, the default mode network allows mind wandering to begin, which can lead to interesting impressions and associations. I remember staring at the edge of each tiny cell, separated by a hair's breadth of gleaming white, both seeming to vibrate together in a great cathedral-like mesh of surface effects, such as we see with the intricacy of a flower's veins or the moon's rays spotlighting the rhythmic diffraction of the waves.

For Corballis, the kind of mind wandering that is akin to play is evolutionarily necessary because it is adaptive: it allows us to simplify and 'prepare for life in a complex world. But play itself adds to the complexity, creating a feedback system that enhances our need for further creative play' (Corballis 2014, 14). Other studies which try to explain why default mode network activity is necessary suggest mind wandering could have mental health benefits to do with relieving stress. Wang et al. (2015), Franklin et al. (2013) and Lerner and Witztum (2015) report the benefits of relaxation and mind wandering while looking at landscape paintings. Kaplan (1995) suggests that natural scenes allow individuals to rest, as opposed to the strain of analysing urban and technological structures and associations, and are therefore 'restorative'.[26] Meanwhile, Zajonc (1980) suggested that artists intuit growth patterns in nature, which are repeated in their visual imagery and practice, or they may be explicitly aware of employing such patterns.[27] It also seems important that art introduces novelty, or even 'noise', into life, breaking habits or clichés that have become robotic and mind deadening.[28]

Christoff et al. view mind wandering as fluctuations in attention to external stimuli, from little or no attention in dream states to spontaneous mind wandering: 'While hiking on a forest trail, a woman's thoughts move from the gravel on the path in front of her to a slug crawling up a stump, and then to a leaf floating in a puddle' (Christoff et al. 2016, 723). In such examples, mind wandering and spontaneous thought occur *with* external objects, as if the objects are players on the stage of mind wandering. What is difficult to disentangle is the way in which 'external players' act as a medium for internally oriented thought. From the philosophical perspective, this is very Heideggerian insofar as Heidegger describes this less as a state of mind and more as a state of being.

Vessel et al. (2012) and Starr (2013) have shown fMRI activation in areas associated with the default mode network in subjects viewing paintings that they reported as beautiful. The authors suggest this is because of the self-referential aspect of the DMN, which requires some self-referential processing and 'self-relevance'. They hold that DMN activation in viewing artworks emphasises the consistency of self-relevance rather than the inconsistences of mind wandering, as do I. This may indicate a presupposition that art must produce harmony rather than dissonance, in which case my approach, diverging from the emphasis given to the ego or self, is not meant to oppose Vessel et al. but to insist on the diversity of experiences that artworks are able to encourage.

---

[26] Further studies by Purcell et al. (1994) of 200 participants from Italy and Australia confirm this preference for landscapes (with water scoring highest); most participants understood nature to be associated with less evidence of human control or intervention.

[27] For further studies, see Kemp (2000) and Wilson (2010).

[28] Perhaps the neuroscientific complement to this idea is that the stochastic aspect of neural processes, commonly understood as noise, is 'essential for maintaining the health of neurons, and it provides the unstructured pre- and post-synaptic co-activity that is required to form new attractors with Hebbian learning, instead of merely reinforcing existing attractors' (Freeman 2000, 217).

In any case, there is much debate as to what might be contained in 'self-relevance'. As I have argued elsewhere (Minissale 2013, 233–250), the self is a complex philosophical and psychological problem. It could be understood in a rich sense (a reflective self, involved in metacognition) or as a kind of vague background hum (a pre-reflective self) or, strangely, as a sense of 'losing oneself'. Vessel et al. accept that

[t]rait studies may reflect a set of processes whereby observers don't simply think about themselves, but, more specifically, match traits with self-inspection, as a part of broader social cognition. In a similar manner, release from deactivation during aesthetic experience may reflect observers' matching self-inspection with their perception of an object (Vessel et al. 2012, 9).

This point complements the idea that internal and external contents can be intercalated with the help of the DMN. What is important in Vessel et al. is the 'intense feeling' evoked by artworks, recruiting aspects of the DMN, and how this might be squared with the mind wandering which also recruits aspects (possibly different ones) of the DMN. We do know that mind wandering and the highly self-conscious thought associated with metacognition are distinct.[29] This allows us to infer 'self-wandering' or wandering ideas about the self, rather than building complex higher-order cogitations on the self, which would reduce the time given to other kinds of engagement with an artwork. Many artworks activate feelings of empathy and ego-dissolution comparable with the 'oceanic feeling' Ehrenzweig describes, where the self is projected onto or finds expression through the features of artworks, even if they are abstract and inanimate. This chimes in with the statement in Vessel et al. that 'our brains detect a certain "harmony" between the external world and our internal representation of the self – allowing the two systems to co-activate, interact, influence and reshape each other' (Vessel et al. 2013, 7). This suggests that the encounter with an artwork recruiting the DMN need not involve an explicit self-awareness and may invoke a relatively formless feeling of transcendent or contemplative oneness with the work (often interpreted as spiritual or otherworldly). This feeling can still activate the markers for 'self-reference'. In other words, contemplation of a Rothko painting may lead to thoughts and feelings which we would cognitively call 'beauty' but these feelings could also involve being in the world, immersion, 'abstraction', or being absent in the world and losing one's sense of self in the grander scheme of things – and in all these examples the DMN could be active.

---

[29] Schooler et al. write that metacognition is obstructed in mind wandering: 'our persistent failure to catch ourselves mind wandering could occur because mind wandering occupies the precise brain regions that are necessary for noticing it' (Schooler et al. 2011, 323).

I suggest, along with Starr (2013, 62), that this external–internal interplay may well be supported by the default mode network. Mind wandering in particular seems to be an example of this, where the chaotic dynamics cued by the low-level perception of a landscape allow rhythmic switching between different mental states. This switching or interweaving between 'internal' and 'external' contents would feel indistinguishable, particularly if accompanied by emotional or affective dimensions: empathy arising from motor imagery modelled by the rhythm of the landscape, the sinuous reach of trees, winding river beds. Balzac reminds us of a character who finds 'a mysterious relationship between his emotions and the ocean's movements … to guess matter's thoughts' (Bachelard [1942] 1999, 172). Bachelard takes up the outlines of this idea and suggests that 'an *extraordinary* tempest is a tempest seen by someone who is in an *extraordinary* psychological state' (172). Bachelard deepens this reverie, noting that correspondences 'are established in rare and solemn moments. Inward meditation provides contemplation in which the innermost recesses of the world are disclosed. *Meditation with closed eyes and contemplation with wide-open eyes suddenly have the same life*' (173, my italics).

One feels sleepy looking out of the window of a train or car to see rolling hills, clouds, houses, trees and telegraph wires in topological freefall. Objects appear purely as notations of rhythm, more or less complicated patterns on a horizontal plane, alternating with blank intervals of open fields. Each of these variable granularities and patterns, dense or scattered, dark or light, high or low, may be accompanied by memories and emotions which rise and fall with this differentiation/de-differentiation. Patterns, shapes and forms absent-mindedly observed during mind wandering can be implicitly understood to be beautiful. But one would risk breaking off from one's immersion in the landscape to describe it in this manner. For such a word powerfully composes the whole view. Indeed, it may well be in this moment that self/other and internal/external distinctions come into view, when the scene is being 'objectified', making the objectifier present before fragmenting again into uncomposed matter in motion with mind wandering. Viewers often find the visual aspect of dynamic fragmentation pleasurable, as we see with fireworks, confetti, polka dots and fountains. Is this an aesthetics or an ontology of nonlinear dynamic complexity?

Georg Northoff, a specialist on the brain regions associated with sustaining a sense of self (cortical midline structures or CMS), raises interesting questions about the default mode network and suggests that 'the brain itself, the resting state's intrinsic features, may provide an input yet to be explored specifically in relation to the neural processing of extrinsic stimuli' (Northoff 2012, 358). Taking a lead from Kant's notion of the transcendental ego, Northoff suggests that the resting state imposes a kind of order on external stimuli, even though it is more often associated with capriciousness, and this can happen when the resting state's 'self-specific organisation is assigned to the stimulus in different degrees' (358). On this view, the self can be instantiated in terms of amounts. Part of, or a degree of, the self can be appended to a part or degree of the external stimuli identified. This implies a variable many-to-many relationship between internal and external resources, rather than one-to-one. In terms of art, one could imagine this happening when we feel that part of a painting speaks to us in very personal ways. This does not necessarily mean that the whole rich sense of self is activated, nor does it warrant a return to the traditional view that art is meant to service the ego or self in an act of the will. Rather, it suggests that art can appeal to 'bits of ourselves'. We see this in examples of Dada collage, where the artwork is able to arrange an uncanny experience of a self in fragments, mingling with other fragments of objects or other selves. Seen in this way, art that stimulates daydreaming and odd juxtapositions disorients automatic processes of identification and even self-identification. The notion of a self that arises by ordering and unifying the great variety of stimuli, sensations and events arising in cooperation with the artwork, as a basis for unified consciousness, is quite different from Ehrenzweig lauding the possibilities of the decomposition of the ego, Joyce's splinters of language, or the fragments in Deleuze and Guattari's 'schizoanalysis' where the schizophrenic is described as being 'as close as possible to matter, to the burning, living centre of matter' (Deleuze and Guattari 1983, 26). It is by virtue of the schizophrenic's suspension of rationality and coherence that these connections to matter can be sustained. A gentler and less alarming ego-dissolution is provided by Csikszentmihalyi, referring to a rock climber: 'One tends to get immersed in what's going on around him, in the rock, in the moves that are involved ... so involved that he might lose consciousness of his own identity and melt into the rock' (Csikszentmihalyi 1990, vii).

In Lewis Carroll's *Alice in Wonderland* (1865), 'the self' becomes so many fantastical animal, vegetable and mineral parts in a rich exploration of the imagination, which we can associate with the daydreaming of the default mode network. The links between the DMN and external perturbations are not arranged by reason to reference a preformed, unified sense of self taken off the peg. It is not surprising that the Surrealists loved this book, because it appears to defer rational control and allows jumbled features of categories to trigger sensations and uncover fragments of lost dreams, memories or desires. These fragments are improvised to form new arrangements which may arise as vague feelings or presentiments rather than fully formed concepts. This might even happen with something as abstract as a sinuous line or a braided surface that seems familiar from previous mind-wandering episodes and contexts.

In any case, talk of brain regions and networks needs to be related to the rather more variable and unpredictable aspects of neural oscillatory dynamics. Based on results from various experiments, Northoff suggests that

internal thoughts in particular, and possibly external contents in general, are more related to low-frequency oscillations whereas cognitive processing of external stimuli instantiates higher frequency oscillations. Hence, it may be the balance between high- and low-frequency oscillations that sustains the balance between internally and externally oriented attention and thus between internal and external contents (Northoff 2012, 385).

Though persuasive, this does not explain what happens when frequencies co-occur and what qualitative experiences are supported by frequencies that are neither high nor low. Meanwhile, De Bruin and Kästner (2011) suggest a dynamic, looping or repeating process of decoupling and recoupling, with external details shifting to thoughts about internal states, gradually and with degrees of self-organisation. This is complicated by the fact that some minimum attention to an internal state can be co-present while being attentive to external objects and, in the reverse direction, one can have a minimum awareness of external objects while attending to internal states (Dixon et al. 2014). The idea that it can only be one or the other seems rather simplistic.

Schooler et al. (2011) emphasise that external contents are 'dampened' in order to concentrate on internal processes of thought, but not altogether ignored. This is very important for artworks such as a Rothko painting, where the dampening effect is already achieved in the external world. In such situations one is able to introspect with the artwork, rather than by turning away from it. This suggests that observation of matter painting involves a predominantly slower frequency (delta or theta) rather than the higher frequencies (gamma or beta) associated with perceptual binding, analytical focus and analysis of form. Yet the cooperation of the two bands could result in a 'feeling for' matter in the process of becoming diffuse or, in the opposite direction, becoming self-organised. This 'feeling for' depends on avoiding the kind of metacognition that insists on strong distinctions of internal and external, 'self' and 'other', 'me' and 'it'.

The idea that the perception of external things can cooperate with mind wandering rather than mind wandering being a withdrawal into

Figure 2.8 Diagram published in Cees Van Leeuwen and D. J. A. Smit, 'Restless Minds, Wandering Brains', in S. Edelman et al., *Being in Time. Dynamical Models of Phenomenal Experience* (2012), 122. Published by John Benjamins Publishing Company, Amsterdam/Philadelphia.

introspection is examined in Van Leeuwen and Smit (2012). The authors suggest that 'perceptual multistability' involves mind wandering; a wandering mind can discover several ways of seeing a figure. The example they use is a geometric pattern.

They write that mind wandering can 'discover several structures in this figure, such as the star (lower right part of the figure), a staircase on the main diagonal, and a house (in the upper right)' (Van Leeuwen and Smit 2012, 125), and switching from one configuration to another is rhythmic and semi-chaotic (Figure 2.8). For Van Leeuwen and Smit, mind wandering plays a key role in 'perceptual dynamics' in which 'stability and instability of neural synchronization patterns are tightly interwoven' (125). This suggests that chaotic dynamics work on the micro and meso levels of neural oscillations, switching from gamma/beta (deliberately focused thoughts requiring narrow band frequencies) to delta waves involving mind wandering and less planned and more unfocused 'broadband' gazing. This can be envisaged as panning over a larger 'landscape' of percepts rhythmically and superficially without fixating – one can think of the reveries produced by observing clouds, with all the transformations that could occur by switching focus from the general to the particular and back again.

Fine-grained vision is understood by Van Leeuwen and Smit as 'analytical vision', while coarse-grained vision is referred to as 'holistic processing': 'With holistic processing, attention is "spilling over" to neighbouring regions' (Van Leeuwen and Smit 2012, 132) when mind wandering takes place. For them, dynamic systems theory is best suited to give an account of interactions between fine-grained vision and holistic processing (124). It seems that perceptual mind wandering is not stochastic but chaotic, following preferred yet complex patterns (with alpha and theta phase synchronisation) across brain regions, as if trying to send out feelers in an itinerant manner to find distant connections. This would require being open to suggestion, not being driven by a purposeful analytical vision that seeks patterns in the world in order to match them to a preconceived idea held in the mind. Knowing what one is looking for considerably narrows this openness. These chaotic dynamics can quickly change direction or amplitude, depending on contingencies, but also allow order to emerge as a phase transition when something is singled out in the scanning of the visual field: 'Once observers engage in a [focused] task, the large-scale patterns of alpha synchronization tend to disappear' (Van Leeuwen and Smit 2012, 138) This complements Freeman's earlier research findings that chaotic dynamics intervene between perceptual states and Tsuda's chaotic itinerancy between attractor ruins. Interestingly, it is this kind of perceptual multistability that would be operating in the famous example of Leonardo da Vinci viewing stains on the wall, where battle scenes and other shapes and forms rise up in the imagination or fall away. This is often accompanied by an oceanic feeling of relaxation and mind wandering, spontaneous shape formation with external content.

The approach taken by Van Leeuwen and Smit goes against a strict separation of mind wandering into introspection and focused attention to external contents, proposing that mind wandering is possible while looking at external things in the form of perceptual multistability. One criticism of this approach is that mind wandering seems to be pegged at the level of perceptual nonlinear switching, without reference to conceptual or emotional mind wandering, which are other levels that can run parallel with or further elaborate on the perceptual level. It would be a mistake to think that the diagram produced by Van Leeuwen and Smit functions only at the perceptual level as a purely analytical puzzle. Perceptual fluidity also involves relaxing eyes, focus and concentration, yielding to rhythmic ambiguities, which produce remote associations, divergent thinking, memories, pulses of sensation and affects. This unfocused perceptual and conceptual mind wandering is undoubtedly what is sustained while viewing Jean Dubuffet's work, as well as Antoni Tàpies's paintings in the early 1950s, which were often described as walls. The photography of Brassaï was well known for its focus on graffiti and made a profound impact on Tàpies. De-differentiation liberates us from dwelling too long on one particular aspect and allows us to see the whole view, the inherent instability of matter, and to see it not as forms but as fluctuations, or 'portraits of energy', as Bachelard would say in his works on reverie.

Many artists, designers and art historians train themselves to look in 'unlikely' places and create 'wholes' that are improbable, imaginatively joining up parts of images to create abstract patterns where battles emerge from stains on a wall. This scanning or panning activity is metastable and open to random associations and gestalts. In terms of neuroscience, Van Leeuwen and Smit suggest that perceptual mind wandering can take the ordered, rational processes of analysing a scene offline:

Mind-wandering could break up the existing pattern structure, such that the components can be reassigned to a different pattern . . . . Items in the background – or, if you like, outside of the focus of our attention – tend to receive weaker groupings than those in the foreground. Thus, when the mind wanders away from a certain grouping, it will dissolve and release its components for novel use (Van Leeuwen and Smit 2012, 123).

2.4   Mind Wandering and Creativity

Studies in creativity indicate that the executive and default mode networks can cooperate in dynamic alternation or can co-occur. Golland et al. (2007) examined subjects watching films and found activation in brain areas associated with stimulus-driven activity as well as in cortical regions whose activity was dissociated from the external stimulation. They suggest that this is an intrinsic system working in parallel while watching films, 'largely overlapping with the task-negative, default-mode network, comprising areas associated with – as yet not fully understood – intrinsically oriented functions' (Golland et al. 2007, 766). Although Hasson et al. (2004) found intersubject brain synchronisation patterns in film viewing, commenting on these findings Golland et al. pointedly remark on the correlated activity of the two networks, externally and internally focused, that 'embedded within this correlated territory, there were several anatomical "islands" that consistently failed to show inter-subject correlations' (Golland et al. 2007, 766). The authors assume that the intrinsic, non-intersubject behaviour is associated with *not* processing the film. But this is a logical inference that no longer holds. That the authors found a 'striking' complementarity between the two networks is surely highly significant in suggesting that film viewing creates a complex interlocking of internal and external contents. As Golland et al. state, additional work is required to discover the precise nature of what is happening in this inter-digitation of networks that is so important for our understanding of what occurs in our encounter with artworks. Significantly, in a later study, Simony et al. (2016) find that default mode network activity is shared by individuals given the same stimuli. This suggests the intriguing possibility of 'social dreaming' where individuals' minds wander in similar ways and for the same intervals while listening to narratives and by extension watching films and viewing art.

Studies by Dumas et al. (2010) add intriguing dimensions to the internal/external exchange of information. Simulations indicate intra-individual brain synchronisation favouring the alpha rhythm of relaxation or mind wandering. Kelso et al., commenting on the findings of Dumas et al., note that, by directly linking intra- and inter-brain synchronisation, the study 'opens a way to draw a connection between neural, behavioral and social scales and thereby extend the general theory' (Kelso et al. 2013, 128). Given that executive control and the default mode network can work together, another research question concerns how a subject's executive control works while viewing a film or artwork that simulates (or is a product of) DMN activity, and whether this would be similar to the subject's own DMN activity. In their study on 'intersubject synchronisation', Hasson et al. discovered shared brain patterns in subjects watching films and report that 'despite the completely free viewing of dynamical, complex scenes, individual brains "tick together" in synchronised spatio-temporal patterns when exposed to the same visual environment' (Hasson et al. 2004, 1635). In particular, this synchronisation is observed while subjects watch 'delicate hand movements during various motor tasks ... faces, buildings, open landscape scenes and ... objects' (1637).

In their interpretation of the Limb and Braun (2008) study of jazz players, Dixon et al. (2014) suggest that externally directed cognition associated with the executive network and internally directed cognition involving the default mode network actually co-occur. This is supported by other findings where researchers had rap musicians perform repetitions of lyrics to create rap freestyle (Liu et al. 2012). Twelve male rappers were studied. Improvised rapping resulted in increased activation of the DMN and decreased activation in the executive network, but there were nevertheless alternations between the two networks during improvisation. Dixon et al. offer an elegant explanation of how the brain distributes attention across internal and external contents:

While engaged in a conversation, attention may be directed externally to the words being spoken by a friend, but simultaneously directed internally to inferences about their mental state and interpretations of the meaning of their words ... while driving to a novel destination; there may be rapid shifts between focusing externally on the road and other cars, and focusing internally to remember the directions (Dixon et al. 2014, 322).[30]

A brain area can be activated during a task but cannot play a critical role in performing the task: 'rather, it might be "listening" to other brain areas that provide the critical computations' (Sawyer 2011, 150). An intriguing possibility is that the default mode network is monitoring external contents in an 'open' but shallow or low-key manner. Openness – not fixating on one specific part of the visual field – could be explained in neuroscience through the 'sentinel hypothesis', according to which the DMN maintains a general low-level focus of attention when, like a sentinel, one monitors the environment for unexpected events. This may be explained in terms of nested or alternating brain frequencies normally associated with distinct or opposing functions (focusing and defocusing, for example). In viewing art, we could be involved in a broad information gathering or sweep, which happens during configural viewing or panning over large areas, but still be able to switch to focal points or anomalies.

Neuroscientific research into eye tracking (Massaro et al. 2012) reveals that the interpretation of figures in art involves top-down (that is, conceptual and perceptual) *a priori* categories and structures, while less organised and more 'abstract' landscape scenes may employ bottom-up (more spontaneous and rhythmic) interactions with low-level perceptual details.[31] This has been termed variously a low-key 'exploratory state' (Shulman et al. 1997). An important part of this may be related to the soothing features of landscapes that artists specialise in isolating and projecting. This may involve recursive eye behaviour affecting, and being affected by, meditative or hypnagogic states, as we see with the fracticality of a flickering fire, or in Pollock's and Still's artworks. Again, it appears that the logical inference that one cannot pay attention to external contents while mind wandering, which is illustrated by Figure 2.9, is too simplistic.

This diagram is described by Buckner in the following way:

---

[30] The authors also explain that 'some aspects of creativity, such as the process of generating novel ideas, is thought to rely on a mind-set characterised by largely spontaneous processing and seems to involve the *co-occurrence* of EDC [externally directed cognition] and IDC [internally directed cognition] with minimal interference' (Dixon et al. 2014, 324).

[31] At the neuroscientific level, studies suggest that 'the generation of novel and creative ideas is accompanied by low arousal of brain-activity and is mediated by inhibition of top-down control' (Shamay-Tsoory et al. 2011, 179). This also involves inhibiting stereotypical automatic thinking. On the other hand, damage to areas responsible for top-down control and self-monitoring and evaluation can cause subjects to have difficulty deciding what is important and how to evaluate decisions and possibilities.

Passive Task   # Active Task

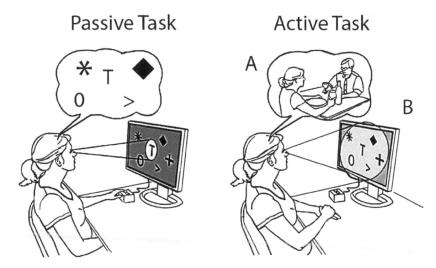

Figure 2.9 Diagram of attentional and mind-wandering states, published in Randy L. Buckner et al. (2008). 'The Brain's Default Network: Anatomy, Function, and Relevance to Disease'. *Annals of the New York Academy of Sciences* 1124 (1): 19. doi:10.1196/annals.1440.011.

As one goes from an active task demanding focused attention (left panel) to a passive task (right panel), there is both a change in mental content (A) and level of attention to the external world (B). Spontaneous thoughts unrelated to the external world increase (A). There is also a shift from focused attention to a diffuse low-level of attention (B) (Buckner 2008, 19).

I am particularly interested in how (A) and (B) work, and many of the studies I have engaged with in this section struggle to find a model that accurately describes this relationship in terms of the underlying neuropsychology. Of course, the diagram on the right, particularly (B), looks like a very dull episode and would not be consistent with the experiential aspects of viewing artworks or films that recruit bottom-up processes. (B) would not explain the reverie involved in observing a fire or viewing examples of abstract art, which may cut across the default mode/executive control dualism. We can think beyond the dualistic assumption with the sentinel hypothesis and by showing how mind wandering can be maintained below the level of consciousness that we characterise as cognitive or voluntary vision. This is not far from the old psychoanalytical notion that conscious thought and subconscious thought do not cancel each other out but are co-present. Mind wandering can occur as a complementary non-rational exploratory activity while conscious viewing *seems to be* dominant, and this complements Krauss's intuitions about the optical unconscious. Mrazek et al. (2014, 236) suggest that a series of *partial* co-activations of mindfulness (executive control) and mind wandering (DMN) occurs, rather than the full-on activation of both brain states, which would be difficult to reconcile. Thus mind wandering and DMN activity would continue to resonate in the background of mindfulness (perhaps as an attractor ruin) or, in the reverse direction, mindfulness could be 'listening in' on mind-wandering episodes or in lucid dreams. This complements the idea that events from the past can be processed non-consciously while more mindful activity is occurring (Hasson et al. 2015, 304). On this view, activity would proceed on multiple timescales throughout the brain, co-activating slower and faster frequencies in different areas at different times. Again, nonlinear dynamics seems the optimum way to describe this complexity.

Danijela Kulezic-Wilson describes nonlinearity in music as minimalist and repetitive structural principles producing trance-like effects. We see this in composers like La Monte Young and Terry Riley, both of whom aimed for 'mystical experiences associated with "expanded time" as opposed to clock-measured time' (Kulezic-Wilson 2015, 99), which was the result of various influences from Eastern philosophy and art. Kulezic-Wilson goes on to explain that the equivalent of this nonlinearity in film is

created through extended duration in the type of cinema known as contemplative, or Slow Cinema, which is generally perceived as an art-house opposition to goal-oriented, dialectical, linear-narrative cinema. The aim of a sense of temporal stasis in contemplative cinema is often motivated by similar aesthetic, philosophical and spiritual concerns as in minimalist music, justifying the use of the term 'transcendental style' (Kulezic-Wilson 2015, 99).

Many of the sequences employed in such films provide the pleasure of experiencing abstract patterns that are ambiguous and formless, continually in flux. The delay in constructing narrative or meaning allows remote associations to be made unexpectedly, even though partial recall of the faded forms or cadences of dreams seem to appear involuntarily and through improvisation. Again, it is the DMN that is associated with improvising unusual combinations of ideas, and perhaps certain film sequences and works of art allow this process to be sustained as part of a creative process. But it may also be the case that the mind wanders when things in the environment are not stimulating: the subject may have low-level interest for external contents and hence the mind wanders. This can be used by film directors to produce different rhythmic variations, some tightly woven that require willed focus and concentration, others that allow the mind to rest and the viewer to consolidate themes and reflections. Passive scanning of an artwork, allowing the space and time for things to unfold, and creative thought may sometimes be synonymous.

Significantly, Ellamil et al. (2012) observed co-activation of the executive and default networks during creative interpretations of artworks, suggesting that one slips in and out of self-awareness during creative evaluation. The authors conclude:

[T]he dichotomy between generation and evaluation appears to be ubiquitous in psychological theories of the creative process, with novel ideas produced during generative phases and their utility assessed during subsequent evaluative phases. This dichotomy is also present in artists' accounts of their own creative process, which they often describe as alternating between rough sketching of ideas and critiquing those ideas to guide the next cycle of sketching and critiquing (Ellamil et al, 2012, 1783).

This suggests alternation between different networks but, closer to studies of co-occurrence, the authors note that the activity over the two networks is highly correlated throughout the task of drawing on a pad in an fMRI scanner. In all likelihood, this switching from executive and evaluative processing to default associative and affective processing is fluid and probably works in the same way as the coupling, decoupling and recoupling theorised by De Bruin and Kästner (2011), which consists of 'smearing' between states attentive to internal contents and states attentive to external contents. The executive system is useful for checking the relevance of ideas that arise spontaneously in mind-wandering episodes and it may be required in the evaluation of creativity. For this to happen quickly during sketching, drawing or painting, the dynamic coordination of the two networks will be needed.[32] Ellamil et al. cite many studies that demonstrate the role of executive control in various creative endeavours including

piano improvisation, creative story generation, word association, divergent thinking, fluid analogy formation, insight problem solving and visual art design. During these creative tasks, high cognitive control enables a deliberate, analytic mode of information processing that facilitates the evaluation of the utility of novel ideas and allows individuals to focus on the pertinent task details and to select the relevant generated ideas. Therefore, the executive network may contribute specifically to the evaluative mode of creative thought (Ellamil et al. 2012, 1784).

The authors note that while 'evaluating the products of one's own creative activity, creative individuals frequently pay attention to their "gut reactions"' (Ellamil et al. 2012, 1784). They conclude:

---

[32] For consistent evidence of the dynamic interactions between default mode and executive control networks, see Beaty and Jung (2018).

On the basis of the previously reviewed neuroscientific findings, we could hypothesize that: (i) the memory network may contribute to associative processes that would enable creative generation; (ii) the default network may contribute either to creative generation through its role in low cognitive control or to creative evaluation through its role in affective and visceral evaluative processing; and (iii) the executive network may contribute to the analytical evaluative processes required during creative evaluation (Ellamil et al. 2012, 1784).

Baird and Smallwood suggest that 'mind wandering enhances creativity by increasing unconscious associative processing' (Baird and Smallwood 2012, 1121), which adds an important dimension to the assumption in Ellamil et al. of a largely conscious creative process. The default mode network is also helpful in creating random or remote associations, something we see in Surrealism, where divergent ideas (for example, the fur cup in Oppenheim's work) come together to stimulate strong affective and emotional responses.

Automatic writing is an obvious example of this process of alternation between networks. Automatism requires cultivating an attitude of passive receptiveness, low-key perception, relaxation, letting the mind wander or be distracted, while we are scribbling with a pen or pencil (as we often are on the telephone when we doodle). After this, the mind can become more evaluative about the shapes and forms we have sketched, turning them into larger or more complex designs or ideas. It seems important that, in viewing automatist sketches, subjects can remain open to possible solutions rather than committing to any one of them. This is the hallmark of 'divergent thinking'. Highly important for Surrealist practice, divergent thinking involves the ability to resist premature closure in order to generate unusual juxtapositions and combinations of ideas.

In aesthetic theory, this phenomenon was studied by Umberto Eco in *The Open Work* (1989). According to this influential approach, the viewer does not simply receive from the artist a communication of a state of mind; rather, the viewer receives a communication of a field of possibilities – of 'equiprobability', or what I have been calling metastability – and the open work, the artwork, presents such a balance of order and disorder that this metastability is offered without closure. Eco compares this phenomenon with Dubuffet's paintings that appear to be made of gravel, stains and dirt, put in a frame and hung on a wall. Eco writes that such matter paintings are 'much like a road surface or other bare terrain in their attempt to reproduce absolute freedom and unlimited suggestiveness of brute matter' (Eco 1989, 90). Eco elevates the freedom and unlimited equipossibility involved in aesthetic experience to the realm of the highest ethics, for he maintains that openness 'is the guarantee of a particularly rich kind of pleasure that our civilization pursues as one of its most precious values, since every aspect of our culture invites us to conceive, feel, and thus see the world as possibility' (104).

2.5 Metastability and Emotion

Up to this point, I have discussed the nonlinear dynamics of the brain in terms of neural assemblages, oscillations and frequencies, perceptual multistability, mind wandering and creativity. The absence of any discussion on how emotion plays a part in this dynamic complexity might suggest that such experiences are cold and detached, but this is not the case. In fact recent studies show that conceptual processing, rather than being an optional supplement to emotions, helps to constitute them. Conceptual categories (e.g. genre, place, type of scene, tradition, culture, script) provide emotion with numerous situational examples (or 'situated conceptualisations', as neurocognitive psychologists call them) without which our emotions could not be experienced meaningfully.[33] Rather than being implemented by a discrete module, emotions appear to result from distributed circuitry throughout the brain: 'Within this distributed circuitry, diverse brain states for a given emotion arise, each corresponding to a different situated conceptualization' (Wilson-Mendenhall et al. 2011, 1109). And rather than having a core set of fixed properties, emotions are continually nuanced by contextual and situational conceptual processing that involves the interplay of memory and the semantic, sensorimotor and nervous systems, each of which in itself does not create 'an emotion'. It is worth noting that 'there is not one bodily signature for each emotion, [and] the same body state across different situations can be conceptualized as different emotions, depending on the situated conceptualization active to interpret it' (1108).

In experiencing art, we habitually speak of how artworks are able to stimulate certain emotions and sensations. Yet, compatible with these recent studies, the particularities of each artwork help to provide and configure a number of 'situational conceptualisations' which give these emotions (and their sequencing, layering or overlapping) a particular, emergent dynamics. Artworks enable multiple transient and ambiguous emotional complexes to emerge in a chaotic manner. In other words, in viewing abstract art, 'the situation' does not have to act as a structuring and defining device for precise emotional responses. It is the emotional dimension of a mind-wandering state, which can be sustained in watching films, listening to music or observing matter in art, that seems important in mitigating the excesses of executive control used to extract a meaningful message from the artwork.

---

[33] See Barrett (2006), Barrett et al. (2009) and Wilson-Mendenhall et al. (2011).

Developmental neuroscientist Marc Lewis's approach, which has garnered much interest and support, is to adopt nonlinear dynamics to understand the emergence of emotion:

I suggest a bridge between emotion theory and neurobiology based on dynamic systems (DS) principles. Nonlinear dynamic systems operate through reciprocal, recursive, and multiple causal processes, offering a language of causality consistent with the flow of activation among neural components. ... Dynamic systems are also characterized by the emergence of wholes out of interacting parts, through processes of self-organization and ... circular causality. Such multilevel causal processes can relate coherent wholes such as appraisals, emotions, and traits to the interaction of lower-order constituents, integrating levels of description both for emotion theory and emotional neurobiology (Lewis 2005, 169–170).

According to Lewis, and as Wilson-Mendenhall et al. (2011) and Pessoa (2010) are at pains to stress, one of the most important problems in the psychology of emotion is that each emotion state is an event that involves many variables from sensation, perception, attention and appraisal to memory and language. All of these affect, and are affected by, various neural dynamics and neurotransmitters. The cognitive aspect of emotion is associated with appraisal, which

denotes an evaluative or interpretive function that is critical for *eliciting* emotion .... If appraisals are necessary to specify emotion states, yet emotion states influence subsequent cognitive processes, then there may be a continuous flux or stream of evaluative events in which appraisal and emotion are interspersed (Lewis 2005, 170, 172).

Lewis shows how it is possible to think of emotion and appraisal in nonlinear bi-directional causal terms, using aspects of dynamic systems theory. For him, cognition can be viewed as self-organising: 'the spontaneous emergence of order from nonlinear interactions among components of a dynamic system' (Lewis 2005, 173). He writes:

The meaning of 'nonlinear' is twofold: (1) cognitive activities are viewed as reciprocal and/or recursive, as characterized by multiple feedback cycles, such that cause-effect relations are bidirectional or multidirectional; (2) effects in such systems are generally not linear functions of causes (Lewis 2005, 173).

Lewis questions the clear demarcation of appraisal and emotion in a cause-and-effect relation, referring instead to an emotion-appraisal amalgam or 'emotion interpretation'. He emphasises that an emotion arises as a global state that is stabilised through numerous transactions of appraisal and emotion based on circular causality. Circular causality is important because it describes how macro and micro levels and 'higher' and 'lower' order functions work together:

A coherent, higher-order form or function causes a particular pattern of coupling among lower-order elements, while this pattern simultaneously causes the higher-order form. The top-down flow of causation may be considered an emergent constraint (by the system as a whole) on the actions of the parts. Cognitive scientists have begun to model higher-order mental states such as intentionality and consciousness as emergent forms that constrain the activation of the psychological or neural constituents producing them (Lewis 2005, 174).

Important in all this are 'trigger events' or 'perturbations' that could be initiated by other individuals, animals, books, films or works of art. It is a nonlinear dynamic or dynamic systems approach that is most effective in conceptualising how various events interact across internal and external resources. For emotion in particular, Lewis maintains that

appraisal-emotion episodes begin with a triggering event. In DS [dynamic systems] terms, the orderly behavior of the system is interrupted by a perturbation, *resulting in a rapid loss of orderliness and an increase in sensitivity to the environment*. The perturbing event must have an impact on one or more system elements or it is invisible to the system. This impact works through a process of nucleation in which the affected element enters into reciprocal, self-enhancing interactions with neighboring elements to create a core proto-organization. Thus, a trigger marks a phase transition, characterized by sudden change and temporary disorder, as the system switches to a new organization (Lewis 2005, 176, my italics).

Art historians would be fascinated by the analysis of mental events in cognitive neuroscience, particularly because it provides hope for interpreting art other than by adopting mechanical and purely reductionist approaches in neuroaesthetics and cognition. For Lewis, the emotion-appraisal complex suggests a fine balance between cognitive processes, on the one hand, and spreading involuntary effects outside of cognitive control on the other. He describes this complex in terms of pathways, and it is interesting that he acknowledges that 'the art of fiction writers and movie directors is to deliberately shape the evolution of such pathways' (Lewis 2005, 176). At the same time, an individual's emotional experience is nested within larger cultural contexts that provide contours for the nucleation of emotional complexes. The situation and the individual will give these larger cultural formations that 'normalise' emotional expression a particular and idiosyncratic twist. The subpersonal is folded into the personal and, in the other direction, the personal is uploaded into cultural traditions. In all this, the most important thing to remember is that the emotion arc is nonlinear and part of a larger non-emotional system.

How does the artwork help to sustain the dynamic interaction between these different networks that modulate attention to external content?

Perhaps art allows us to 'hedge our bets' and court ambiguity through perceptual multistability, which causes the individual to mind wander through several different affective and cognitive states. How does perceptual multistability cooperate with emotional metastability? Are they mutually reinforcing? This is hard to imagine with the unemotional diagrams used to demonstrate perceptual multistability. Art, however, can help to produce emotions that are wave-like or erratic, to accompany perceptual multistability. We see this combination clearly in Pablo Picasso's *Guernica* (1936) or Jean Fautrier's *Otages* series.

Eryilmaz et al. (2011) note the deactivation of several regions of the default mode network while subjects watch strongly emotional films, probably because such films demand more focused attention. But in the resting period after watching the films, DMN activity increases as a way of 'working through' the emotions elicited. The DMN thus has an important part to play in processing emotions. This is supported by Martins and Mather (2016), who show that the DMN is integrated with other brain networks such as the executive network to work through the negative effects of emotions, allowing well-being to emerge. This is particularly so for older adults who have increased connectivity between the different brain networks. Depression is associated with decreased connectivity between the DMN and the different networks, and increased connectivity within the DMN, responsible for periods of introspection and 'rumination'. These studies suggest that the connectivity, toggling or synchronisation between networks, allows for the 'release' and 'working through' of transient emotion episodes. This kind of connectivity between networks can be increased by mindfulness training, which suggests that the skill of toggling between internal and external contents can be acquired.

The chaotic dynamics involved in switching perception states, identified by Freeman and Tsuda, may also have relevance for the switching of emotion states. Switching or toggling between these different ways of emotional engagement with art then becomes a complex internal and external inter-digitation of perception, emotion and cognition, in a feedback loop (not a linear succession) with various mind-wandering episodes that help to elaborate these changes. Artworks trigger emotions and more ambiguous affects and moods that are supported by various oscillatory frequencies across functional networks. If Morse Peckham and other theorists of art are correct in asserting that one of art's more important functions is to provide aspects of chaos, it is able to do so by creating and supporting the condition of metastability, the chaotic dynamics between settled emotion and perceptual states that is often ignored in art history and psychology because it falls between the 'resolved' states that are so often hailed by aesthetics. Rather than ordering emotional and cognitive hierarchies, the composition or disintegration that an artwork arranges can help to destructure such an order. It can also, through cognitive dissonance, wreak havoc on neat demarcations of mental states favoured by reductionist paradigms in art and science.

Figure 1.1 Robert Morris, *Untitled* (1968). New York, Museum of Modern Art (MoMA). Felt, asphalt, mirrors, wood, copper tubing, steel cable and lead, 668 × 510.5 cm, variable. Gift of Philip Johnson. 504.1984. Digital image © 2020. The Museum of Modern Art, New York/ Scala, Florence. © Robert Morris/ARS. Copyright Agency, 2020.

Figure 1.2 Francis Bacon, *Study for Portrait, Number IV (from the Life Mask of William Blake)* (1956). Oil on canvas, 61.5 × 51 cm. Digital Image © Museum of Modern Art, New York/Scala, Florence. © The Estate of Francis Bacon. Copyright Agency, 2020.

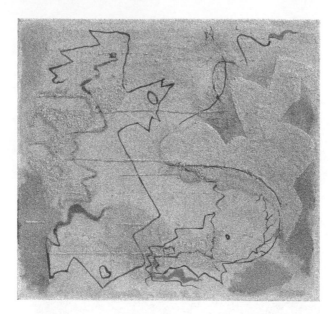

Figure 1.3 André Masson, *Shadows/Les ombres* (1927). Oil and sand on canvas, 50.2 × 54 cm. Digital Image © Christies/Scala, Florence © André Masson/ADAGP. Copyright Agency, 2020.

Figure 2.1 Jackson Pollock, *Guardians of the Secret* (1943). Oil on canvas, 123.19 × 191.77 cm. Digital Image © Museum of Modern Art (SFMOMA), CA, USA. De Agostini Picture Library/Bridgeman Images © Pollock-Krasner Foundation/ARS. Copyright Agency, 2020.

Figure 2.4 Jackson Pollock, *Stenographic Figure* (1942). Oil on linen, 101.6 × 142.2 cm. Mr. and Mrs. Walter Bareiss Fund. 428.1980/ Museum of Modern Art, New York. Digital Image © Scala, Florence © Pollock-Krasner Foundation/ARS. Copyright Agency, 2020.

Figure 3.1 Henri Matisse, *Interior with Aubergines* (1911). Distemper on canvas, 212 × 246 cm. Digital Image © Musée de Peinture et de Sculpture, Grenoble, France. Gift of Henri Matisse, 1922/Scala, Florence © Succession H. Matisse/Copyright Agency, 2020.

Figure 3.3 Salvador Dali, *Autumnal Cannibalism* (1936). Oil on canvas, 65.1 × 65.1 cm. Digital Image: Tate Modern © Fundació Gala-Salvador Dalí, Copyright Agency, 2020.

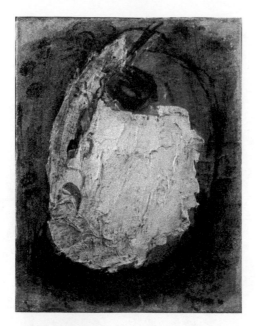

Figure 3.6 Jean Fautrier, *Otage n°7* (1944). Oil on paper mounted on canvas, 35 × 27 cm. Digital Image © Private Collection/Scala, Florence © Jean Fautrier/ADAGP. Copyright Agency, 2020.

Figure 3.7 Jean Dubuffet, *Soul of the Underground (L'ame des sous-sols)* (1959). Oil on aluminium foil on board, 149.6 × 195 cm. Digital Image © Mary Sisler Bequest, Museum of Modern Art, New York, USA/ Bridgeman Images © Jean Dubuffet/ADAGP. Copyright Agency, 2020.

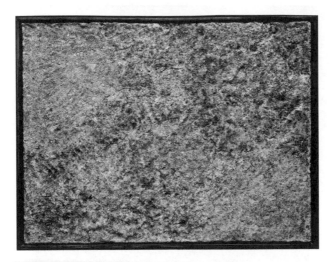

Figure 3.8 Jean Dubuffet, *Place for Awakenings* (1960). Pebbles, sand, and plastic paste on board, 88.4 × 115.2 cm. Digital Image © The Museum of Modern Art, NY. Gift of the artist in honor of Mr. and Mrs. Ralph F. Colin/Scala, Florence © Jean Dubuffet/ADAGP. Copyright Agency, 2020.

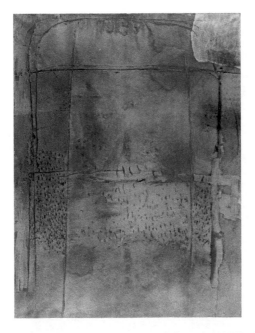

Figure 3.10 Antoni Tàpies, *Ochre gris* (1958). Oil paint, epoxy resin and marble dust on canvas, 260.3 × 194.3 cm. Digital Image: Tate Modern © Fundació Antoni Tàpies, Barcelona/VEGAP. Copyright Agency, 2020.

Figure 3.11 Alberto Burri, *Composizione* (1953). Burlap, thread, synthetic polymer paint, gold leaf, and PVA on black fabric, 86 × 100.4 cm. Digital Image © Peggy Guggenheim Collection, Vercelli/ Scala, Florence © Fondazione Palazzo Albizzini Collezione Burri, Città di Castello. Copyright Agency, 2020.

Figure 3.13 Simon Ingram, *Automata Painting* No. 3 (2004). Acrylic and gesso on canvas, 90 × 90 cm. © Courtesy the artist and The Fletcher Trust Collection.

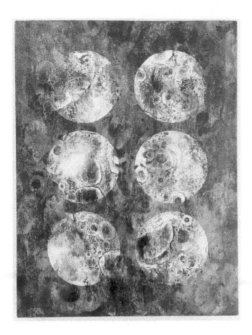

Figure 3.16 Shannon Novak, *Hadrian* (2019). Acrylic ink on board, 25 × 30 cm. © Courtesy the artist.

Figure 3.18 Lynda Benglis, *Contraband* (1969). Pigmented latex. Overall (irregular), 7.6 × 295.3 × 1011.6 cm. Purchased with funds from the Painting and Sculpture Committee and partial gift of John Cheim and Howard Read. Inv. N.: 2008.14. Digital Image © Whitney Museum of American Art/by Scala © Lynda Benglis/ARS. Copyright Agency, 2020.

Figure 3.19 L. N. Tallur, *Unicode* (2011). Bronze and concrete, 183 × 152 × 117 cm. © Courtesy the artist.

(a)

(b)

Figure 3.20 Lee Mingwei, *Guernica in Sand* (2006/2015). Mixed media interactive installation. Sand, wooden island, lighting, 1300 × 643 cm. Installation view: 'Lee Mingwei and His Relations', Taipei Fine Arts Museum, Taipei, 2015. Image: Taipei Fine Arts Museum © Courtesy the artist.

Part III
Rhythm, Dirt, Art

Part III deploys the theories and approaches presented in Parts I and II, along with art historical texts, to develop a new interpretational framework for artworks that make rhythm and matter explicit.

3.1   Emergence

In *Brushstroke and Emergence* (2015), art historian James D. Herbert uses the idea of emergence to interpret art by Courbet, Monet, Cézanne, Picasso and others. One way to explain the concept of emergence is to suggest that an entirely new phenomenon with novel features can arise from the interactions between components or processes which themselves possess none of these novel features. For example, the movements we observe in stock markets come about through numerous small-scale transactions between individuals and businesses, but large-scale economic events, such as exponential growth or collapse, take on qualities and a momentum of their own, independent of the lower levels of activity. Using this conceptual structure, Herbert suggests that neural activity and the movements of the artist's hand, refined or stylised through learned routines and practice, produce brushstrokes that are rhythmically worked up into a greater whole, forming the image of a face or a landscape. In fact, what comes together is the recognisable style of an artist's depiction of a face or landscape, a kind of handwriting that is chromatic and texturally complex. The larger scale, where the structure and character of the brushstrokes are seen to support the image (the face or landscape) and the overall composition, is complicated by cultural learning and social and aesthetic values of which the artist may not be aware. Herbert mentions an ant colony, where each ant acts locally but produces global effects unknown to it, which he compares to the artist whose brushstrokes are built up locally with little or no knowledge of the broader discourses surrounding the work (Herbert 2015, 36).[1]

---

[1] Herbert emphasises the lack of conscious deliberation in brushwork, even though artists often do adapt and respond to other artists' brushwork. He suggests that, in Georges Seurat's *Young Woman Powdering Herself* (1889), the dots seem to play on the idea of the

Herbert sees the emergence of the brushstroke as a discursive element in the late nineteenth century, when Gustave Courbet and Édouard Manet began to make the brushstroke itself more visible or available for discussion in ways that it was not in the 'polished' and refined mirror-like surfaces of Jean-Auguste-Dominique Ingres or Théodore Géricault. Perhaps because in emergence theory novel features of the higher level are not reducible to the lower, only general impressions are offered concerning the cognitive neuroscience and kinaesthetics which act in concert with the application of paint.[2] Instead we have registers that only connect because they share some superficial visual features of 'collectivity' or granularity: 'brushstrokes' is the higher register for the collective action of neurons. Then there are a lot of analogies made using 'collectivity': ant colonies are brushstrokes, or brushstrokes are crowds because they are collectives. Herbert's suggestion that the activity of applying brushstrokes remains spontaneous and psychologically outside deliberate or studied control ignores one important aspect: the oil medium lends itself continually to erasures and reworkings of the brushstrokes. If an artist places too much emphasis here or there, or adds too much of a particular colour in an area where there are many brushstrokes, they will undoubtedly smooth it out, or overpaint it, or correct it in favour of a configural view, so the artist does actually have time to ponder the appropriateness of their brushstrokes.

The fact that a brushstroke may have erased another prior brushstroke has important consequences for the theory of emergence and control:

[1] particles of talc the woman dabs on her body, which illustrates a complex and unpredictable emergence from the local practice of dabbing dots on a surface, brushstrokes like a swarm of ants seeking out a theme (Herbert 2015, 102). This seems to place too much emphasis on spontaneous emergence *during* painting. The 'emergence' could have occurred months earlier in Seurat's mind, before he executed the work, when he was obsessed with the notion of granularity; it could have been the product of several other instances of 'emergence'. Although emergence is loosely explanatory for innovation, such innovation can quickly become the material for later emergence ad infinitum. Thus it seems important that we try as hard as possible to avoid the perpetual elusiveness of innovation, at least by giving some account of what novel features *emerge from*, even if we avoid reducing them to this lower level.

[2] For example, Herbert speaks of axons and dendrites of neurons and rightly explains that concepts concerning beauty and truth cannot be traced back to these microscopic features (Herbert 2015, 2). However, there are many important intermediate registers, such as brain areas, networks and brain oscillatory frequencies, which are ignored. Moreover, the registers have mutually reinforcing relationships. Where is the emergence to be observed: in all of them together or incrementally at various levels? The registers require more rigorous definition rather than being alluded to metaphorically in general terms. Instead we are left with a number of unfalsifiable propositions about complex phenomena and where they come from. Emergence theory does not mean that one is excused from accurately describing the activity of the lower registers from which these complex phenomena are supposed to emerge. Cellular automata are not brushstrokes.

top-down conceptual control at the 'global' level does not just emerge, but is in constant interaction with bottom-up local contingencies and chance occurrences, which the artist is often happy to accept. Countless times artists leave little accidents and colour contrasts and unexpected gestalts in their work because they are pleasing in the overall design, or they may paint over them. Evaluation and spontaneity alternate or inter-digitate, as psychological studies show (Ellamil et al. 2012), and metasta-bility propagates across different registers, from neurological to perceptual and phenomenological, in far more complex ways than the analogies concerning 'collectives' suggest. Perhaps this is what Herbert is edging towards when he mentions that the agency of the artist could be distributed among 'competing forces across registers' (Herbert 2015, 95). This continual metastability and feedback activity between different registers, from small to large scales and back again, tends to undermine clear distinctions between registers. Emergent features do not just appear at the top level but are distributed across the registers in dynamic loops.

Herbert's attempt to move away from the *cogito* and intention as the organising force ordering each brushstroke complements Krauss's attempt to question conscious artistic intent. Style, for example, could be seen as an emergent feature in the sense that the artist is intent on solving particular perceptual problems or pursuing a particular combin-ation of sensations, which are in themselves semi-chaotic or involuntary, nonlinear and dynamic. This activity may spark reflections about more complex cultural or aesthetic ideas or concepts that may not have been intended at the perceptual level but have the character of being ordered and intelligible. The notion that the artist composes with metastability, with aspects of chaos and order of this kind, makes emergent properties seem less mysterious and marred by explanatory gaps.

An important complication is that, in working on a painting, the artist imagines what it might look like in the first micro-seconds of an encounter with it, the crucial moment when the attention is piqued, or not at all. Given that a painting can take weeks, months or even years to complete, how does the artist go back to this 'state of surprise' except by adopting a complex psychology of avoiding rational analysis? The artist must imagine this encounter by adopting various defamiliarising devices. This could involve covering the canvas, as art historical tales inform us Titian was wont to do, to forget what the painting looks like for several weeks in order to see it anew, as if stumbling upon it accidentally without preconceptions.

Scientists are fond of reminding us of the top-down/bottom-up distinc-tion. Top-down processes are 'cognitive influences and higher-order mental functions ... learned associations' (Kandel 2016, 22), and semantic and cultural associations. Bottom-up processes are understood as the basic automatic non-conscious processes of perception hardwired in the brain, to do with sensations and feature detection. The production and reception of artworks involve important complexifications of these different processes. One of the most interesting things about Paul Cézanne's practice, for example, is the way in which the brushstrokes are distributed to create a dense metastability: a whole view can be maintained in the working memory and can degrade when the brush-strokes are viewed individually to provide a shimmering effect where fine-grained details and larger shapes are in tension with each other. Here, composition and dispersal are unstable; sometimes we turn more to one than to the other, and we enjoy the variable rhythms of doing so. Cézanne's paintings seem to make explicit something that we perhaps normally do while viewing more traditional landscape paintings. Sometimes the leaves in the trees remind us of brushstrokes and some-times brushstrokes are leaves, landscapes or emotional fluctuations. This is also what we do when we look at trees and landscapes in nature, zooming out to gain an overall sense of the main shapes and lines and zooming in to see the finer details, somehow achieving a metastability where the two views seem to be simultaneous (even if one has to squint).

Further dimensions to these ideas are provided by Maurice Merleau-Ponty in his famous essay 'Cézanne's Doubt', where he skilfully teases out the important implications of the philosopher's allusive text:

Cézanne did not think he had to choose between feeling and thought, as if he were deciding between chaos and order. He did not want to separate the stable things which we see and the shifting way in which they appear. He wanted to depict matter as it takes on form, the birth of order through spontaneous organization ... an object in the act of appearing, organizing itself before the eyes (Merleau-Ponty 1993, 63).

To focus on a single brushstroke or patch of colour in a Cézanne painting, ignoring the surrounding multiplicity of marks and groups of marks, would not only be difficult (as one is variably aware of one's own periph-eral vision) but would also break up the configural, shimmering effect, 'the vibration of appearances' (Merleau-Ponty 1993, 68). This shimmer-ing contracts aggregates of details and lets them disperse and expand across the painting. Although it is commonplace for vision to focus on small details and configure them into larger wholes very quickly, the painting slows this process down, making us more aware of these switches in focus. This perceptual shimmering is not just a visual sensation or effect, but may be accompanied by the stirring of an aesthetic emotion.

Again, philosophy finds the language to reveal the full significance of this, as Merleau-Ponty writes:

We live in the midst of man-made objects, among tools, in houses, streets, cities, and most of the time we see them only through the human actions which put them to use. We become used to thinking that all of this exists necessarily and unshakably. Cézanne's painting suspends these habits of thought and reveals the base of inhuman nature upon which man has installed himself (Merleau-Ponty 1993, 67).

Heidegger might suggest that this is a 'letting go' of the world's logic and object percepts that bind materials, and it is through painting that these presentiments emerge. In *Informel*, this matter and sensation without objects are taken even further. In Cézanne, the boundedness of objects is just beginning to be undone. The letting go, which is a kind of emergence of relaxation, is also a letting go that produces a reverie which is both introspective and 'out there' with the rhythmic geologies and ephemeral grasses of the landscape, where we 'join the wandering hands of nature' (Merleau-Ponty 1993, 67). The patterns of matter in the land, whether mud, grass or pond, their transposition into art, and our reverie involved in viewing it share underlying rhythms which we only become aware of in moments of insight. This provides an interesting context for various perceptions of the landscape by poets and writers. Bachelard quotes a passage from André Gide:

It seemed to me that the landscape was but an emanation from myself, a projection, a vibrant part of my own self – or, rather, that I could feel myself only in it, that I was its centre, that all had been dormant, virtual, inert, before my arrival, that I had brought the landscape into being, step by step, in perceiving its harmonies, that its centre of consciousness was in me. And I stepped out, marveling, into the garden of my dreams (quoted in Bachelard [1948] 2002, 295).

Pointedly, Bachelard goes on to say that 'such centres of concentration are naturally not to be plotted out geometrically' (Bachelard [1948] 2002, 295). Gide's words can also be compared with the well-known statement by Cézanne: 'The landscape thinks itself in me and I am its consciousness' (Merleau-Ponty 1993, 67). All of these statements attempt to express how internal introspection and external perception become entwined, transforming the act of observation into participation.

3.2   Henri Matisse

In his essay on Henri Matisse, the art historian Mark Antliff defines 'becoming' as a 'permanent state of coming into being' (Antliff 1999, 185), which he suggests can be a form of creativity. Matisse's striking juxtapositions of isolated patches of strong colour, applied with visible rather than smoothed-out brushstrokes, create the impression of luminosity and heat, causing lines and forms to shimmer. The following is Antliff's description of Matisse's *Harmony in Red* (1908):

Fauvism's colour harmonies assimilated the *vibrato* of matter into the durational consciousness of the artist. ... *Harmony in Red* causes our attention to oscillate between volumetric depth and an unmodulated decorative surface, as if the fluctuating field of red introduced a rhythmic pulse to our reading of the painting (Antliff 1999, 185).

It is interesting that empirical research suggests the colours red and blue encourage different kinds of cognition due to the different ways in which they stimulate the brain's oscillatory frequencies. Red was found to enhance attention to detail, while blue encouraged creative associations between items in a scene (Mehta and Zhu 2009).[3] This is apart from the common understanding in many cultures that red is associated with danger, while blue is a melancholy colour or helps to produce such an affect. Colour, when understood as variable wavelengths of light, is a good example of how matter is transformed and carried across the domains of brain, body and world. Cézanne understood this intuitively and suggested that 'colour is the place where our brain and the universe meet' (Grene 1993, 221). And for Deleuze, the viewer of a painting experiences 'the sensation only by entering the painting' – and, one would expect, by having colour and rhythm 'enter' the spectator – 'by reaching a unity of the sensing and the sensed' (Deleuze 2003, 35).

Antliff's description of the visual and spatial rhythms provided by Matisse's paintings could be supported by another level of description: the viewer becomes aware of many different kinds of ambiguity in figure–ground relationships and the density of design. This is the case with Figure 3.1, *Still Life with Aubergines* (1911), which almost appears to be abstract in nature. The complicated and interlaced patterns create all kinds of perceptual ambiguities, and a strong metastability arises from the alternation of disintegration and composition. This work was inspired by Matisse's encounter with North African Islamic art. The aesthetic can be described in purely rhythmic terms. See, for example, Alexandre Papadopoulo, writing of Persian art, which also features complex chromatic tensions:

[W]hat counts within the autonomous world of each miniature is the exact position of each colour area with its specific decorative elements forming more or less dense, more or less rapid groups, each of which enters the relationship with all the other groups in their separate allotted positions so as to make up a tightly woven nexus of relationships of density, tempo and color.

---

[3] Specifically, they found that red enhances performance in a detail-oriented task, whereas blue enhances performance in a creative task.

Figure 3.1 Henri Matisse, *Interior with Aubergines* (1911). Distemper on canvas, 212 × 246 cm. Digital Image © Musée de Peinture et de Sculpture, Grenoble, France. Gift of Henri Matisse, 1922/Scala, Florence © Succession H. Matisse/Copyright Agency, 2020. (A black and white version of this figure will appear in some formats. For the colour version, please refer to the plate section.)

Each coloured surface, along with the decorative motif overlying it, is caught up in innumerable relationships of this order with all the other coloured surfaces and decorative motifs. This sets up an immense polyphony of topological relationships, which is similar to modern painting (Papadopoulo 1980, 104, 111).

This description participates in nonlinear dynamic thought. It is the result of a way of seeing, and perhaps being, for a while at least, and helps to suggest this to the reader. In *Concerning the Spiritual in Art*, Wassily Kandinsky also wrote that Persian art had a 'complex rhythmic composition, with a strong flavour of the symphonic' (Kandinsky 1977, 34). The nonlinear dynamics of colour sensations interplaying with complex patterns may be felt as a restless, agitated allegro. The blink can also switch to a general frame where one imagines fireworks and constellations. I have already mentioned that a direct relation between internal neurological events and external dynamics at the phenomenal level is evident from the fact that alpha waves (employed in mind wandering and panning over a scene in a nonlinear manner) strongly affect, and are affected by, eye movements. Purposeful analytical thought blocks such waves with a simple attentional blink, often accompanied by a focus on logical questions.

One such question might be: why aubergines? An inference is that Matisse was interested in exploring a range of colours and tones, complementing or contrasting with this purplish-brown base, used as a key, which provides a sense of continuity through many perceptual dynamics involving frames, screens, mirrors and windows. But this explanation is itself prompted by the allure of the earthy rug, the bursting forth of blue petals, the rich and fertile umber outside the window that leads us back to the aubergines. The painting momentarily seems to possess a creative circular causality all its own. The dynamics of different biological sensitivities in the viewer are intercalated with multiple variations in the environment, in the visual field.[4] This is far from the idea that 'I *will* see x, y, z' and closer to 'This is happening to me'.

In explaining Bergson in the context of Matisse, Antliff intuitively understands the main principles involved, describing states of mind as 'varying degrees of rhythmic tension or relaxation roughly corresponding to the degree of freedom inherent in a given activity' (Antliff 1999, 196). This relaxation allows new associations to arise beyond literal description. Importantly, Antliff picks out a description in Bergson, who states that the work of art is able to induce an aesthetic feeling produced by rhythms, which provide certain constraints or cues that can be picked up across different minds: 'The painting therefore is both an integral, organic entity imbued with the rhythmic duration that created it, and an agent for creative development on the part of the beholder' (196).

3.3   Cubism

Linda Dalrymple Henderson (1983) has shown that the Cubists, at the beginning of the twentieth century, were interested in discussions concerning the fourth dimension, with very limited knowledge of Albert Einstein, whose work became better known after Cubism's first flowering. Cubism, particularly of the analytic period, consistently depicted objects by breaking the rules of Illusionism, which stem from common-sense presuppositions about the structure of reality. In Illusionist painting, space was organised in terms of static, bound objects, placed in an ordered, rationally calculated

---

[4] Mondrian remarked that this multiplicity is not just a thought but 'is now visible as such' through his painting (quoted in Golding 2000, 30).

perspective structure. The figure was contrasted with a relatively feature-less ground, and a sense of time was achieved by intuiting the reach between stable objects or even the walking distance between them. These conventions were a way to situate the viewer in front of the proscenium. Even if the Cubists were primarily interested only in exploring innovations in pictorial conventions, their rejection of Illusionism as an aesthetic system also implied the unravelling of the phenomenological coherence that Illusionism was meant to provide. Cubism thus encouraged different ways to think about and experience space, time and matter as constituents of reality. And in any case, Bergson had prepared the ground with his hugely popular lectures at the Collège de France that fed the imagination and ensured the circulation of new ideas about matter and mind, the *élan vital*, and duration. Timothy Mitchell summarises how Bergson's ideas influenced the pictorial experiments of Cubism:

The old differentiations of art: figure and ground, solid and void, static and mobile, are unified by a flux of a homogeneous medium. Being is replaced by becoming. The distinctions between space and matter are no longer visible. The uniformity of brush stroke and surface texture unite the entire canvas. A strong sense of a common substance, now fluid, now solid, comprises the whole (Mitchell 1977, 181).

The 'window on reality' of academic painting established an ideal rela-tionship between the image and the viewer, a relationship that stabilises the phenomenology of observation as a kind of linear, rational inspection of objects from an inertial frame. Here, the viewer becomes master of the proscenium assembled before them and judges the success of the work in creating the illusion of depth and following the conventions of compos-ition. In such painting, landscapes tend to have their horizons and transi-tions of colour and form; portraits have identifiable people; and still lifes have their solids. These paintings have their front and back, their above and below. They arrange convincing solids in intervals of space, often in coded relationships to establish intelligible rhythms and correspondence. Cubism suggested that objects were no longer solid – even the structure of the painting and the proscenium no longer seemed secure. The sugges-tion of a fleeting world was unsettling, or exciting, and transformed the viewer's failure to grasp the object into new perceptions about the reality of contingency, partial vision and ephemeral minutiae.

In Georges Braque's *Violin and Palette* (1909), the violin seems to shift its position, showing us different facets of its cube-like form repeated in space, seeming to pass through phases from past to present to future. This undermines the presupposition that the viewer is at a fixed point of obser-vation. What the viewer sees is either a set of objects that are not in a stable position (the Illusionist tools to indicate this have been withheld), or the viewer intuits that they themselves are shifting positions in front of the picture – or worse, that both picture and viewer are shifting vis-à-vis each other. This serves to stimulate a broader metastability where the sense of order and stability is thrown into disarray. The painting appears to anticipate the viewer's shifts of position while moving itself, and no matter how often this 'trick' is viewed, again and again it erodes stability delightfully.

In Figure 3.2, *Violin and Palette*, Braque paints an Illusionist device, the nail casting a shadow at the top of the painting on which a palette has been

Figure 3.2 Georges Braque, *Violin and Palette* (1909). Oil on canvas, 91 × 42.8 cm. Digital Image © Solomon R. Guggenheim Museum, New York, USA/Bridgeman Images © Mondadori Portfolio/Walter Mori/ Copyright Agency, 2020.

hung. This creates a dissonance with the Cubist mode of representation. At the top we have a 'fixed' object, the nail, and below it the vibrating sheet of music and the violin evoke the intangible continuities of time and sound. Cubism thus suggested the contingencies of vision, not in terms of the fluctuating light of Impressionism but through the undermining of common-sense Euclidean certainties and one-point perspective. In many Cubist paintings, objects are depicted as undergoing change or being in motion, and the viewer's own eye movements, nonlinear and iterative, add to the overall impression of a dynamic encounter.

As if to underline how painting can reflect shifting paradigms in what is seeable and sayable, Cubist painting suggested a wholly different under-standing of reality underneath the visible and sensible, so that viewers could imagine fluctuations of energy (movement, reflections of light) from several spatio-temporal vantage points, and even from the flurry of brushstrokes 'leaking out' across the boundaries of objects. This took the very act of observation away from certainty and to a place of doubt.[5] Carl Einstein, a critic writing around the time of Cubism's first exhibitions, notes that the experience of Cubism is not merely a theory but leads to a 'gradual modification of sensations' in which 'a person waxes and wanes ... in the sensation of himself or his feeling for objects, in the harnessing of time', so that what is represented is 'the very history of the sensations, experiences brought close up, whose symptoms are at best so-called things' (quoted in Haxthausen 2011). As Haxthausen writes:

What is represented is not, as in illusionistic art, ostensibly stable objects that exist apart from perceiving subjects, but rather our own unfolding subjective process of vision as we apprehend and mentally construct the volume of objects in moving toward and around them in space. In short, a cubist painting presents us with a synchronic image of a diachronic process, a process Einstein called, *das vor-stellende Sehen*, envisaging seeing (Haxthausen 2011).

A kind of gestalt process of gathering form and organisation occurs when one studies how Braque composed the painting to suggest that the

---

[5] This understanding of the psychological effects of Cubism seems at odds with realist functional interpretations which propose that Cubism allowed the percept of the object to be retrieved from the pieces, fragments and cues found in the work. This is how Cubism has been interpreted using simple logic, for example, by Zeki (1999), but there still remains the involuntary engagement with flow, texture and falling apart, which is rhythmic and sensuous rather than simply a function of object recognition. It is interesting that this intuitive engagement with Cubism actually complements the more scientific view that objects are only ephemeral collections of particles undergoing change. As with the rabbit-duck illusion, it is possible that in viewing a Cubist painting some viewers see flux and instability while others join up the dots to see the object they believe is intended, while there are some who are happy to keep flipping the perceptions with increasing skill, so that a certain transitional state is viewed or imagined.

palette, which is pinned to the wall by the nail, is the head of the
musician tilted to the left. But it is an unstable and continually disintegrating composition that appears to be animated by jerky multi-
directional movements which we imagine might accompany
a musician's playing. As Deleuze writes in *Bergsonism*, 'multiplicity'
has had two fates in the twentieth century: the first is phenomenology,
where multiple phenomena are unified in conscious experience, and
the second is 'Bergsonism', which aims to maintain the heterogeneity
and simultaneity of different perceptual and material states (Deleuze
1991, 115–118). This is seen as a positive and creative involvement in
the spontaneity of material forces, before the 'grasping' of concepts and
percepts binds such forces into ordinary tools for everyday use. What
Cubism and particle physics, or atomism, have in common, despite the
fact that many if not all of the latter's theories may have been unknown
to the Cubists, is that both undermine naive sureties about visual phenomenology and its automatic identification of solid objects and their
fixed locations. For their different reasons, both Cubism and particle
physics suggested that ordinary, everyday vision is not all there is. For
particle physics, the 'ontology' of objects is in the eye of the beholder and
their temporal frame. Objects are the interweaving of different uncountable particles of energy, perceived as mass in a certain place and time by
an observer. Imagining a non-object kind of physicality can support
a different kind of time frame. For Cubism, experimenting with new
ways of depiction included breaking the logical rules of presenting
objects in order to spark novel psychological effects, namely impressions
of shimmering energy not contained by objects, a curious time delay in
their 'solidification', with incongruent haptic triggers – jazz rather than
melody. The nail, the palette, the violin scroll, various visible brushstrokes, all provide stable attractors amidst the perceptual jolts. The
mind wanders between focused and unfocused views. The configural
view suggests various simultaneous twitches and convulsions and an all-
round rhythmic agitation that can be read as forces exerted from within
the figure or as pressure applied from without.

While particle physics may rely on a mathematical or statistical understanding of the nonlinear dynamics of energy exchanges, art is content
with suggesting the experiential quality of nonlinear dynamics through an
appeal to the perceptual ambiguities of texture, edge detectors, transparent or opaque forms, and mental rotation. This results in a game of hide-
and-seek with various occlusions, all sustained by a dense patterning of
non-rational geometries. This leads us to the level of description concerned with the formal and rhythmic qualities of Cubist paintings. As the
art historian David Sylvester writes, insights can occur 'while releasing

oneself to the rhythms of art' (Sylvester 1997, 335). He goes on to describe various artists in rhythmic terms:

Differences are that the pace of the flux in Braque tends to range from adagio to andante, in De Kooning from allegretto to presto, that the things in Braque's flux tend to be things that can be held in the hand, seen indoors, or taken indoors, whereas the imagery in De Kooning's flux is wider, more elemental, wilder (Sylvester 1997, 335).

In Braque's *Violin and Palette*, the angles of the blocks are illogical and irregular. Staccato rhythms gather pace with the fine-grained texture created by the dark stippled brushstrokes and cross-hatching in the corners of these blocks. The brushstrokes are applied uniformly in horizontal or diagonal lines, cross-hatched on the palette and zigzagging in the curtains. This increases the feeling of complex movement and the switching of angles. The artist not only arranges complex rhythms and counter-rhythms, based on facets of cubes that seem to twist away from each other, but also allows the matter – the paint itself – to remain visible with rough, thick and unblended variations of impasto. The warm brown tones we associate with the violin give way to shades of green, which could refer to the velvet of the curtain. Many of the cube-like forms are highlighted with predominantly quicksilver and pewter tones, suggesting reflections of light (and on closer inspection some of these areas of impasto do actually catch the light). The violin seems 'staggered' in different phases of time, not just where it is but also where it was and where it will be, and the violin strings suggest different chords or pulses of sound across space and time.

There is a nonlinear dynamic of reading this kind of art: fragments of autobiographical memory that we use to summon identity, violin scrolls, palette, curtain and musician dissolve into the ruffled surface. Meanwhile, each block or cube can be 'read' sequentially, for nested within each area is a localised pattern of brushstrokes that is rough or smooth, uniformly packed or erratic. Michael Thaut ventures to say that one of the fundamental mechanisms at work in producing and listening to music is simultaneity and sequentiality and that music allows one to appreciate both at once (Thaut 2005, 3). Thaut offers an interesting description: 'In rhythm we hear not just time marked off in the pulsating ticks of a stopwatch; we become aware that time exists in patterns that shape and unfold horizontally and sequentially in interaction with patterns of vertical simultaneity' (15). The visual equivalent of this is holistic or configural seeing, a coarse-grained grouping of parts into a whole view (as we see when we imagine the palette to be the head of the musician), while at the same time being able to appreciate the detail in peripheral vision. The result is a tension between sequential vision attracted by the parts and simultaneous vision putting all the information into a whole. On Thaut's view, this is an aspect of music, the rhythmic multistability of part and whole. But there is also a higher-order conceptual perspective provided by the Heideggerian 'shimmering' of the matter of art, the earth and the recognisable forms that the world excavates from it. In both cases we are dealing with a kind of rhythmic vibration of concepts (a violin, a palette), with sensations of wood and viscosity, which seem to flesh out or even overrun these concepts. We are left with this metastability somewhere between grasping the object in conscious perception and releasing it into a general mind wandering *with* the volatility of matter's surfaces.

In his survey of art curators, the psychologist Mihaly Csikszentmihalyi picks out a particularly relevant quote from someone trying to explain how Cubism is not just an image of geometrical complications but also presents material differences:

I responded to the painting because of its color and forms, they were musical forms; it's called *Guitar on a Table*, and it had all the strength and beauty of a perfect Cubist picture. It also had all of the variety of paint manipulation that you associate with Cubist paintings. Some dry, chalky lines across the surface, that were just laid on as if with a piece of chalk. Other stuccolike surfaces where ashes or sand might have been added with paint to make it really crusty, and other areas of dead black, where you think the artist has collaged a piece of paper to it, it is so flat . . . . Across the face of that painting were many surfaces. There were thin dry surfaces. There were lusciously worked areas in the painting. There were thinly washed areas that were not dry, that still had a shine to them that allowed a transparency looking through to a certain depth within the painting (Csikszentmihalyi 1990, 32).

For me, this account describes not only a painting but a way of seeing and a style of allowing the mind to wander with features in the world rather than ignoring them. Thomson et al. write that mind wandering reveals a tendency for passive watchfulness, engaging with the world in an 'open' attentional state, 'much in the way that practised meditators can train themselves to do wilfully' (Thomson et al. 2015, 189). They note that 'a propensity towards mind wandering may be symptomatic of a tendency to prevent over-investment of attention in any one event, a cognitive style that supports "fluid" interactions with a world that is inherently complex and dynamic' (181). Cubist and Futurist paintings make us acutely aware of multiple processes in motion. One can even reach philosophical concepts by way of this participation in the continual flux of the universe. The Futurist artist Umberto Boccioni fully understood the implications of this when he quoted from Bergson's *Matter and Memory* (1896) that any division of matter into independent bodies 'with determined outlines is artificial' (Petrie 1974, 144). For him, only changing states exist and matter itself is 'envisaged as an indivisible whole ... a flux rather than a thing' (144).

3.4   Surrealism

Although characterised as analytical or intellectual, Cubism could also be understood and experienced simply as bizarre. The Surrealist André Breton claimed Picasso as 'one of us' (Breton 1925, 26). Cubism, much like Surrealism, broke with 'tangible entities ... and the facile connotations of their everyday appearance' (29). But what Surrealism introduced into the crystalline refractions of Cubism and the muscle memory and athleticism of Futurism was something altogether different: it was cheese. I am referring to Salvador Dali's notorious response to the scientist Ilya Prigogine that the artist's wilting clocks were not illustrations of relativity but Camembert melting in the sun. As is typical of Dali, the artist foregrounds not the intelligible matter of science but matter that is *bassesse* (lowness), unleashing a smearing of dissonance with consonance: sticky, messy, viscous, erotic, unctuous, toxic and yet edible. While Cubism created fragments out of objects and bodies in a kind of inorganic puzzle or game, matter in Surrealism suggested bodily processes, fluids and secretions, smells and textures, releasing upon the lucid mind feelings of disgust, eroticism, anxiety and the oneiric. Dali channels expenditures of energy (affects) that are not necessarily pursued or processed cognitively or through indirect symbolism, but through direct contact with the rhythms and textures of matter and their assault on the olfactory sense. Imposing rational understanding on his stretched-out, entropic fleshy forms provides some interruption of haptic and motor projections, warding off visceral reactions while viewing these paintings. It is not by any means a minor contribution to the history of thought that Surrealism was able to stage visually the failure of aesthetic distance to repress 'dirty' matter. Many Surrealist paintings present a challenge to the viewer: relax and surrender to stray feelings and presentiments, or wilfully repress these feelings in favour of aesthetic cleanliness. In the Surrealist imagery of biomorphism and metamorphosis, matter is experienced as organic and elastic, affective and oneiric, suggesting a rhythmic, topological process of bending and stretching beyond clearly demarcated boundaries.[6] Because

---

[6] Simondon uses an analogy to explain the difference between classical geometry and topology (Simondon 1964, 52). The mechanical saw works its way through the wood in accordance with measurements and geometry, cutting across the grain and fibres which are organically twisted with each other as knots and growth patterns. The carpenter or sculptor either works with these in terms of rhythm and flux or cuts across them in terms of

the material quality of elasticity can go back and forth (contract and expand) simultaneously in multiple dimensions, it acts as a metaphor for a spasm, an involuntary reflex or convulsion, while the undulating boundaries of flaccid objects, wilting clocks and jagged rocks, suggest a pathetic fallacy where mind wandering dissolves the solid structures of rational thought, becoming the hills and clouds, exposed to desert mirages, seeking refuge in dark porticos. A good summary of this way of 'thinking through' solids is provided by Michel Leiris's description of Miró's *Portraits* of 1929: they capture 'this liquefaction, this implacable evaporation of structures ... this flaccid leaking away of substance that makes everything – us, our ideas and the ambience in which we live – like jellyfish or octopi' (Leiris, quoted in Bois and Krauss 1996, 54). Similarly, the image of thought Einstein used for understanding space-time was a mollusc, a sluggish and slippery jelly able to quiver in several areas and directions at the same time. This imagery is a metaphor for a holistic understanding of space, time and matter – not as consecutive or separate things, but joined in a configural movement and in continual modulation, a simultaneity of local and general energies that can be viewed from a distance as a quivering thing. It takes a special kind of non-rational imagination, and aspects of mind wandering, to 'see' this topological (nonlinear) dynamics in the mind's eye. If the Cartesian grid exemplifies ordered, rational and sequential thought, then visual examples of topology, such as in marbling, cloud forms and strata, which we also see in Dali's images, become visual equivalents of non-rational thought, nonlinear mind wandering and daydreaming. Such topological thinking can be understood as counterintuitive and non-Euclidean, while the Mobius strip or an Escher drawing helps us to keep the complex configural movement and continual modulation of space-time in our mind as a visual aid. This is one of the reasons why matter paintings, and photographs of geological and meteorological phenomena, can help to produce reverie, because they excite an intuitive interest in the viscous consistency of matter and time, half-oneiric, half-scientific.

The logical interpretation of Figure 3.3, Dali's *Autumnal Cannibalism* (1936), concerns the Spanish Civil War (1936–1939), where the bodies eating each other symbolise internecine strife and the country seems literally to be devouring itself. The apple above the lumbering bodies is supposed to refer to the legend of William Tell, in which a father shoots an apple placed over his son's head. The dining room in a desert becomes a strange battle-field of the senses, percepts twisting and turning with indeterminate

---

[6] form. Implicit here is a mode of practice and a psychology that flows with the matter cooperatively and allows it to be preserved or eliminates its existence through function.

Figure 3.3 Salvador Dali, *Autumnal Cannibalism* (1936). Oil on canvas, 65.1 × 65.1 cm. Digital Image: Tate Modern © Fundació Gala-Salvador Dalí, Copyright Agency, 2020. (A black and white version of this figure will appear in some formats. For the colour version, please refer to the plate section.)

boundaries: to the far right, hanging from the drawer of the dresser, is a dark ochre cloth, which, astonishingly, becomes a monstrous gargoyle spewing forth the drawer or even the whole dresser or, paradoxically, appears to devour it, a kind of ouroboros effect. The circularity here is not just a binocular rivalry but a convulsion, a nausea, where vomiting or devouring, desiring or destroying, has equal weight, a theme taken up in the main part of the painting. Dali's clinical Illusionist technique supports haptic vision and muscle memory, with undulating rounded rhythms, smooth and ripe surfaces, firm and resistant yet yielding. These rhythms are interrupted here and there with images that repel and worry: the sharp glint of a knife cutting into soft flesh, a flaccid tongue nailed to the drawer, curious ants clustered around a tear in the skin. The olive bodies are meant to be understood as elastic, textured, simulating biting, cutting and chewing actions, squeezing and twisting, to present an overall view of multiple simultaneous directions and stresses. It is also possible to read the 'language' of the hands. Not only do each of the hands and various fingers produce a formal rhythmic variation across the middle register of the painting, they also indicate an oval shape that can be seen as the hands of a clock. While the hand with a fork, aimed at a shadowy eye at twelve o'clock, seems determined and workmanlike, around three o'clock another hand limply holds a knife and seems perfunctory in its carving. The hand holding a spoon on the end of an arm in the middle of the painting seems to be affected by finicky bourgeois manners, an effete gesture that disguises any suggestion of being motivated by appetite. Meanwhile, the hand below it squeezes what seems to be a breast on a dish. In this way, the hands not only tell the time but also seem to indicate different degrees of the intensity of will, appetite and desire. As the artist forms reveries with the hands, the depictions of the various hands form reveries for the viewer. Their gestures reveal a hunger for flesh as food, alternating with eroticised flesh, both supporting the theme of cannibalism: the eyeless figures devour each other and enjoy each other, 'objectifying self-enjoyment' as the art historian Wilhelm Worringer would say. Viewers will notice different aspects in different order and create different percepts, and some may not even notice the yawning drawer. I have mentioned that each hand seems animated by its own rhythm of intent, each a different kind of mental state, sensation or attitude, each a movement-percept in the complex array of motor imagery routines available. While the bodies are interlocked, propped up by a crutch or tenderly draped over each other in drunkenness, seen configurally these structural details suggest an elaborate psychological machinery of push and pull, pressures and counterpressures, which Freud continually used as a metaphor for describing the drives and the intricate interplay between id, ego and superego.

Although the matter in matter painting is immobile and inanimate, we can easily imagine what it might be like to touch, whether it is rough, smooth, wet or sharp, whether it is delicate and fragile or hard and rubbery. In Surrealism these tactile qualities, where we imagine our hands touching textures and surfaces, are distorted by the irrational tactile qualities of dreams. Bachelard writes:

The eye itself, pure vision, becomes tired of looking at solids. It needs to dream of deforming. If sight really accepts the freedom of dreams, everything melts in a living intuition. The 'soft watches' of Salvador Dali stretch and drain from a corner of a table. They live in a sticky space-time (Bachelard [1942] 1999, 106).

Bachelard notes that the hand 'also has its dreams and its hypotheses. It helps us to understand matter in its inmost being. Therefore it helps us to dream of it' (Bachelard [1942] 1999, 106). Touch is nonlinear in that its procedures are not successive; it tarries in circular movements or strikes out on random paths and in random directions, registering the sustained and continuous chaos of texture while forming mental images. It shares properties with an oneiric duration or automatism because it 'does not have the different resting places provided by successive stages that contemplation finds in working with solids' (108).

In Dali's work the 'topological' intuition of matter serves to meld sensory modalities in rhythmic ways. This pulsative, or convulsive, experience leaks into our more intellectual, detached understanding of the painting as an allegory of civil war. We feel our way into various qualities: the fleshy and erotic, slimy and palpitating, abject and putrescent, flabby or firm, engorged or atrophied, and these attributes clash with the surgical and sterile Illusionist technique with its suggestion of glassy surfaces, a magnifying lens where we discern intricate patterns of light that seem to animate the swarming of ants or the viscous decomposition stretching into the velvet shadows.

The imagination of matter, which is arranged by the contours and rhythms of fear and desire, is strangely echoed in Jean-Paul Sartre's description of the viscous in *Being and Nothingness* (1943), where the subject is attracted to, yet disgusted by, clingy substances, a struggle with desire characterised as an uncomfortable stickiness between the fingers, a rumination that is literally interdigitated with matter. For Bachelard, viscosity can be evidence of an 'oneiric fatigue' that prevents the dream from continuing (Bachelard [1942] 1999, 105). For Bergson, duration has a 'thickness', which he calls *l'épaisseur de durée*. And for Herman Melville, in a more positive vein, daydreaming with the hands and with materials produces not only a thickness of time suspended somewhere between wholly subconscious and conscious domains, but also a certain erotic decomposition of self-awareness handling viscous matter, very different from Sartre's anxieties. A striking passage in *Moby Dick* (1851) describes the process of squeezing spermaceti:

After having my hands in it for only a few minutes, my fingers felt like eels, and began, as it were, to serpentine and spiralize. As I sat there at my ease, cross-legged on the deck; after the bitter exertion at the windlass; under a blue tranquil sky; the ship under indolent sail, and gliding so serenely along; as I bathed my hands among those soft, gentle globules of infiltrated tissues, woven almost within the hour; as they richly broke to my fingers, and discharged all their opulence, like fully ripe grapes their wine; as I snuffed up that uncontaminated aroma, literally and truly, like the smell of spring violets; I declare to you, that for the time I lived as in a musky meadow (Melville [1851] 2007, 464).

For Max Ernst, matter could suggest fugacious and pliable dream images, shapes and forms that shift between desire and fear. He famously adopted various techniques including frottage, which involves placing paper on an uneven surface such as a tree bark and rubbing it with crayon or pastel to allow the uneven grain and pattern to emerge. He also used decalcomania, where thick quantities of oil paint in various hues are smeared onto paper and pressed onto another piece of paper or canvas and then peeled off to reveal matter in an arbitrary and complex state. Ridges and kaleidoscope colours could 'suggest' lines of development, possible shapes which could be teased into larger forms that would emerge out of the chaos, providing new possibilities for mythological beasts and recollections of oneiric motor imagery with their distinct rhythms.

This process involved the kind of reverie prompted by stains on the wall that we remember from Leonardo da Vinci's notebooks. Interestingly, Krauss mentions this example of gazing at stains as 'a latency that permits the welling up of associations with the creative process' (Krauss 1993, 65). This suggests that creativity should not just be identified with the active principle of doing something to the world. It also involves a kind of passive receptiveness open to shape shifting, as well as *Gelassenheit*, the willing of non-willing, the intention to be, at least for part of the time, intention-less. As Sylvester aptly suggests in his description of Ernst's frottage or decalcomania, the mind 'is open to the highest knowledge when it ceases to be an agent and becomes a kind of medium', which underlines the 'need for an inspired passivity as the proper condition for artistic creation' (Sylvester 1997, 203). Ernst writes about a work titled *Napoleon in the Wilderness* (1941): 'Plants turn into living animals; architectural shapes turn into statues which are at once plant, human shape and tropaion ... it is impossible to make out whether a living substance has been petrified or an inanimate one brought to life' (quoted in Polcari 1991, 25).[7] The unravelling of perceptual binding that is responsible for the identification of commonplace objects occurs because the matter in matter painting is too complex to grasp and transform into intelligible forms. This continual deferral of binding may be experienced as an all-over dynamism, as writhing shapes and forms that seem to have a life of their own.

Many Surrealist artworks do not respect the boundaries surrounding objects. Experimenting with oneiric images of mind becoming matter, they seem to provide visualisations of pantheist or panpsychist ideas, except that such perspectives are usually argued in earnest using logic. Particularly in their material experiments, Surrealist artists explored

---

[7] The tropaion is an assemblage of weapons and defensive devices of war, cobbled together to form a monument to victory, the emergence of a resolution. Polcari's description of these Surrealist techniques suggests they aimed for a 'nonlinear, evocative, indirect, unforeseen and multiple associational image' (Polcari 1991, 27).

techniques of composition which could lead to involuntary glimpses of marvellous hybrids stretched and smeared across the animal, vegetable and mineral worlds. These metamorphoses arose in places shaded from the glare of rational inspection. Shadows consoled the ancients when producing their artworks, and it is not surprising that Surrealist and abstract artists felt that this shelter was by no means a modern phenomenon. Making art was not just a question of rationally observing and copying objects in nature or in older art; it also involved experiencing haptic, kinaesthetic and affective rhythms recalled from the dream world and captured historically in structures of mythmaking. For Surrealists familiar with Carl Jung's ideas about the collective unconscious, it was a short step to extend this notion to the collective historical unconscious. Drawing the contours of figures from ancient mythopoetic images was the discovery of one's own unconscious rhythms as much as it was delving into the rhythms experienced by the original artist in producing the work.

The artistic practice of autonomism, or automatic writing, which can lead to metamorphosis or biomorphism, suggests that the rational or common-sense understanding of figures and objects is superficial, lying on the surface of a deeper metastability. Smithson expresses this point with some clarity:

Separate 'things,' 'forms,' 'objects,' 'shapes,' etc., with beginnings and endings are mere convenient fictions: there is only an uncertain disintegrating order that transcends the limits of rational separations. The fictions erected in the eroding time stream are apt to be swamped at any moment. The brain itself resembles an eroded rock from which ideas and ideals leak (Smithson 1996, 112).

Smithson's 'panpsychism' is of a different stripe than Surrealism's. Although literary, it is closer to a monism shorn of the dreamlike qualities of Surrealist metamorphosis, and it substitutes deep time and geology for myth and the oneiric. Smithson, one feels, would have been more at home with the 'neutral monism' of the philosopher Bertrand Russell, who held that reality is neither matter nor mind and arises from a deeper level from which both are derived.

For the Surrealist George Bataille, the meeting of matter and mind is non-cognitive, visceral, rhythmic and startling: 'Love and life appear to be separate only because everything on earth is broken apart by vibrations of various amplitudes and durations' (Bataille 1985, 7). This convulsive level of reality is continually repressed by good manners and ideals to do with the human will:

The vicissitudes of organs, the profusion of stomachs, larynxes, and brains traversing innumerable animal species and individuals, carries the imagination along in an ebb and flow it does not willingly follow . . . . Man willingly imagines himself to be like the god Neptune, stilling his own waves, with majesty; nevertheless, the bellowing waves of the viscera, in more or less incessant inflation and upheaval, brusquely put an end to his dignity (Bataille 1985, 20).

For Bataille, archaeology and anthropology insist on rational categories, classifications and chronologies, which add authority to modern bourgeois values concerning race, social class, the moral order and the origins of man. This ordered system is different from the polyrhythmic, Dionysian, affective underbelly of the imagination, where the forms excavated by the professional archaeologist or anthropologist are considered to be products of dreams. In order to see them this way, an irrational dream logic and underlying rhythms are required. This is Bachelard's phenomenology of affectivity, which 'fabricates objective beings out of phantoms that are projected by reverie . . . it creates images out of desires, material experiences out of somatic experiences' (Bachelard [1938] 1987, 37–38). The museum label fades into the background when we view, for example, a clay jug through the mobility of motor images; swelling bellies, slender necks and pouting lips irresistibly morph into birds and fish. Seen in this way, many other museum objects, with their shapes and forms halfway between quotidian use and the haptic exploration of dreams, can lead to complex cycles of reverie. They also yield to erotic fantasies, the lyrical and mythopoetical, or traces of all of these in nonlinear sequence as we walk round these objects. As Bataille suggests, 'the fact that reason denies any valid content in a mythological series is the condition of its most significant value' (Bataille 1985, 1).

The zoomorphic antiquities which excited the Surrealists provide a non-rational imaginative space of oneiric elasticity which seems to share the aura of the archaic, interiorised into present rhythmic feeling. This is not the 'logical' explanation that mirror neurons are for understanding how another's body is strained by a heavy weight, or when it is weary, but the illogical leaps of the imagination that allow one to feel what it is like to lumber like a strange hybrid animal, swing from a tree, coil like a snake around an arm. We do a lot of this kind of thing in our dreams, but not in a disembodied way: our sensorimotor reflexes are active in dreaming, as they would be while watching a film. These reflexes and projections are functional in dreams of flying or rhythmic transformations into bizarre hybrid forms. Artworks recall some of these rhythms and strange juxtapositions, allowing us to daydream with eyes wide open, following the contours of the artwork.

It is Bachelard who reminds us of the rhythmic dimensions of an oneiric topology, which becomes savage in the writing of Lautréamont. Bachelard singled out Lautréamont for his animal aggression, the sensations and motor imagery. This aggression is also evident in Bataille's writing where the target of this great animosity is not just the bourgeoisie but also rational, dry and descriptive accounts of artefacts such as we read on museum labels and in the typologies of species, peoples and cultures. The rhythmic repetitions of this linear narrative are appropriated by Bataille, who uses this pseudo-scientific style of writing but fills it with Surrealist content: irrationality, hybridity and obscenity.[8] See, for example, his treatment of botany:

[E]ven the most beautiful flowers are spoiled in their centers by hairy sexual organs. Thus the interior of a rose does not at all correspond to its exterior beauty; if one tears off all of the corolla's petals, all that remains is a rather sordid tuft . . . . But even more than by the filth of its organs, the flower is betrayed by the fragility of its corolla: thus, far from answering the demands of human ideas, it is the sign of their failure. In fact, after a very short period of glory the marvellous corolla rots indecently in the sun, thus becoming, for the plant, a garish withering. Risen from the stench of the manure pile – even though it seemed for a moment to have escaped it in a flight of angelic and lyrical purity – the flower seems to relapse abruptly into its original squalor (Bataille 1985, 13).

Bataille introduces a dynamic complexity of sex and flowers, manure and poetry, decay and desire, into the rational order of scientific writing as a kind of sabotage of intellectual housekeeping where everything has its proper place. Deadpan description is arrived at through measured tones adopted 'freely, like a beast of burden, to accomplish ends which are not its own' (Bataille 1985, 81).

One of these ends is to explore Bataille's concept of *bassesse* (base or low materialism), which revels in the low, vulgar and disgusting in order to attack Hegelian rationalism and forms of logical positivism and order. The photographic works of Jacques-André Boiffard, for example, are known for relishing feet, dirt, filthy fingernails, and he produced various pictures of open mouths and secretions, as well as dead flies, for Bataille's Surrealist magazine *Documents*. Another subversive concept, *informe* (formlessness), was intended to unravel rational systems by allowing material spontaneity, aimlessness and anarchy into thought as a source of creative wandering. As Krauss writes, Bataille likened *informe* to '*crachat*' (French for 'spittle'):

[N]oxious in its physical formlessness . . . the removal of all those boundaries by which concepts organise reality, dividing it up into little packages of sense, limiting it by giving it what Bataille calls 'mathematical frock-coats,' a phrase

---

[8] Bataille was also interested in a great many multistable images and reproduced photographs of ancient gnostic medals where animals and humans create hybrids that 'embody' the dualism of idealism/base materialism.

that points both to the abstractness of concepts and to the prissiness with which they are meant to constrain (Krauss 1985, 37).

What I have suggested about the viscous matter enacted by various Surrealist artists finds its textual equivalent in Bataille's *informe*.

The key art historical work that discusses how Bataille's notion of the *informe* influenced artistic practices is Yve-Alain Bois and Rosalind Krauss's *Formless* (1997). The book catalogues some lesser-known 'formless' works such as Lucio Fontana's *Ceramica Spaziale* (1949) and Gordon Matta-Clark's *Land of Milk and Honey* and *Photo-Fry* (both 1969), as well as artists interested in the idea of entropy such as Claes Oldenberg, Robert Smithson and Robert Morris. The repeat structures of Edward Ruscha are also supposed to be formless because they suggest non-differentiation (even though they are very 'formal' in terms of being photographs and solid spatial structures). Krauss considers Cindy Sherman's photographs of formless puddles of vomit and detritus as a parody of misogyny that associates menstrual blood, fleshiness, vomit, bodily fluids with female biology (Bois and Krauss 1997, 240). There is also a muddying of the waters between *informe* and base materialism, involving an interesting treatment of scatological images (114–115, 246–248) with a focus on Cy Twombly and Mike Kelley.

Bataille's *informe* is interpreted by Krauss as a way to debunk modernist myths of pure opticality, detached analytics and technical semiotics, particularly by emphasising the 'pulsative' aspect where non-conscious impulses torque the space-time of rational observation. Thus we have an excellent chapter on the pulsative dimension of the unconscious, based on a quote from Jacques Lacan: 'the pulsative function . . . of the unconscious' (Bois and Krauss 1997, 161). While we are fully conscious, looking out of the window, we may not be aware of the rhythms of the leaves swaying in all directions but this does not mean that these movements are not registering somatically, providing many possible pathways to conscious action. With matter paintings we can surrender to the sensations of rhythm and, in the other direction, when we think we are perfectly in cognitive control, our perceptions are gently guided by less-than-controlled impulses. Krauss writes cogently that high art regulates, aestheticises and emasculates the pulsative: '[I]f the pulse were to enter painting at all, it could only be through the highly controlled mediated rhythms of formal proportion . . . . It is, on the contrary, through the lowest and most vulgar cultural forms that the visual is daily invaded by the pulsatile' (164). The pulsative (or 'pulsatile', in Lacan) 'incites irruption of the carnal' (32). The emergence of this raw rhythm-matter in the artwork draws in its wake the pulse of the unconscious as it buoys our experience of looking. Krauss also talks about the 'oscillating alternation' between high and low, Eros and Thanatos (69). For Bois and Krauss, it is in the disappearance of the first person in viewing and feeling that the formless actually begins, insofar as it is not used as a tool for conscious manipulation seeking to express *my* will.

Bataille writes that 'base matter is external and foreign to ideal human aspirations, and it refuses to allow itself to be reduced to the great onto-logical machines resulting from these aspirations' (Bataille 1985, 50–51). Thus matter does not have to be philosophically elevated or 'pure'. In art, subverting rational control through mind wandering opens one up to the risk of being submerged in 'unclean' thoughts, triggering repulsion, yet this may also be fascinating. This ambivalence is based on an attraction to the rhythms of matter, with repulsion arising from their origins and associ-ations, which is exactly what Andres Serrano takes to its logical conclusion with his photographic cataloguing of animal excrement in close-up. A form of abstraction persists in surrendering to the sensation arising from quan-tities of patterned and striated matter that resembles rocks and sediment, and this conflicts with our 'realist' identification of animal excreta, accom-panied by vague fears of infection and pollution. Base materialism refuses to assume that matter is noble because it is free. For Bataille, 'heteroge-neous matter' is so repulsive that it resists 'not only the idealism of Christians, Hegelians and Surrealists, but even the conceptual edifice-building of traditional materialists' (Stoekl 1985, xi) – and, I would add, latter-day 'new materialists', monist-vitalists and animists.

Bataille's visceral and scatological attacks on reason strongly point up the various characterisations of the matter painting of post-war abstrac-tion in Europe and the United States.[9] Although both were heirs to Surrealist automatism, metamorphosis, biomorphism and chance, American Abstract Expressionism seemed to take a more idealist 'ana-gogic' path to abstraction with Pollock, Still and Newman, whose art Stephen Polcari and others understood as a pure spiritual struggle or

---

[9] This base materialism has also become a dominant tradition in contemporary art. See Santiago Sierra's *21 Anthropometric Modules* (2005–2006), consisting of large brown sculp-tural blocks: the exhibition reads, 'made of human faeces by the people of Sulabh International, India'. The excrement was collected by the low-caste poor who are supposed to atone for their misdeeds in previous incarnations and who dispose of human waste for a living. At this juncture, one is reminded of Hany Armanious's *Year of the Pig Sty* (2007), a kind of dirty ecosystem where mud is recycled or smeared into, onto and under various objects, across categories – shoes, floors, tools and little 'truffles', suggesting scatological remnants and marks. Of course, there are a great many other works of art that reference faeces, from Piero Manzoni's *Artist's Shit* (1961) and Andres Serrano's series of close-up photographs of animal excrement to Chris Offili's elephant dung and Willem Delvoye's elaborate machinery processing organic material to produce waste. It is this kind of work in the public domain that appears to test not only the very limits of art but also any residue of idealism, or panpsychism and vitalism, that elevates any or all kinds of matter.

renewal.[10] This is quite different from Bataille's description of the mole that 'hollows out chambers in a decomposed soil repugnant to the delicate nose of the utopians' (Bataille 1985, 35) or, as Francis Ponge writes, Jean Fautrier's 'dirty' paste work imagined as the product of a 'cat who goes in embers', as in its 'manner of excreting: in pasty, adhesive mortar' (Ponge 2002, 49).

3.5   Abstraction

Even in traditional figurative art there are moments of abstraction. In early Flemish painting, for example, there is a childlike fascination for the folds of drapery gathering around the feet of figures; sometimes neat triangular folds seem to twist and turn, to shape a certain embodied engagement which is arrhythmic and suggests the inner agitation of the figures. This is where abstract thought is allowed a brief duration in the world of Illusionism. The folds we see in Gian Lorenzo Bernini's *Ecstasy of St. Teresa* (1647–1652) seem to arrest the attention more emphatically as part of the main narrative, the ripples in the petrified cloth causing spikes in attention, generating erratic saccades and suggesting fibrillation. These descriptions of expressive values in the observation of inanimate matter, even in works prized for their Illusionist techniques, may be intellectual and detached but there are levels of participation that involve feeling the rhythms of motor imagery and the dazzle and excitement of competing stimuli. On the neural level, this can be understood as a criticality of various neural frequencies that compose or disperse quickly in cooperation with optical stimuli. This provides a full picture of Aby Warburg's intuitions in his notion of *Pathosformel*, which suggested that flowing drapery could reveal inner emotions (Freedberg 2006, 28).

There is of course a detailed history in German aesthetics dealing with empathy, starting with Robert Vischer's coining of the term *Einfühlung* ('feeling into') in the nineteenth century. Controversy over how to define the properties of 'empathy', the term used by Edward Bradford Tichener in 1909 to approximate Vischer's concept (which loses some of the power of the original), has dogged its use and continues today. The term *Einfühlung* was far from being agreed in the exacting circumstances of nineteenth-century German aesthetics. Some art historians, such as Heinrich Wölfflin, intended it as a cerebral association of ideas, whereas Vischer used the term to describe a sensorimotor (and possibly mystical)

---

[10] See Anfam (2007, 51–66) for a discussion on the tropes of the old European art of morbid pessimism and the new American art of spiritual renewal played out in terms of transatlantic rivalries.

involvement with form. These different emphases theorise empathy as either a sensory or a more conceptual involvement with the formal properties of art. There still remains disagreement between psychologists regarding the functional aspects of empathy, whether, for example, to define it as a cognitive phenomenon rather than simply an affective state. Lamm et al. (2007) emphasise that affective states are mediated by cognitive and motivational processes. This suggests that degrees of empathy can extend to movement-matter as well as a body, an animal, a piece of furniture or a cloth fluttering in the breeze.

An insightful study of the effects of German aesthetics and early twentieth-century biological discoveries concerning sensation and emotion is Kirsty Martin's *Modernism and the Rhythms of Sympathy* (2013). Martin writes that Walter Pater and Vernon Lee believed that because humans are 'made up of the same elements that compose the external world ... we are somehow in sympathy with things outside us – and when we become most vividly aware of our own physicality we also become aware of our kinship with the rest of the physical world' (Martin 2013, 50). One way in which this kinship can be explained is that motor imagery is projected onto objects and things even though they are immobile and inanimate, prompted by the relations of lines, curves and the proportions of structure or form, which was also appreciated by Ruskin, Pater and Lee:

The idea of rhythm is crucial throughout Lee's work. Her sense of the shapes of visual art, of the energetic sympathy induced by listening to music, of our response to the movement of lines in the nonhuman, and of our response to the energy of nature, all drew on a sense of rhythm: of how energy might be patterned. Lee considered rhythm to be the grounding principle in all emotion, and in all acts of understanding another's emotion (Martin 2013, 50).

Influenced by Henry Head, Lee also understood that

because our feelings and our life are bound up with energy and the shaping of energy, we are thus able to relate emotionally to patterns of movement and energy outside ourselves. This is most obvious for Lee in music, where she sees our forms of energy as almost embodied in the rhythms of music (Martin 2013, 54).

It seems that Lee's thought about energy and rhythm feeds into higher-order concepts about empathy and art. Rhythm is pursued somewhat more single-mindedly by Hans Kaufmann, writing in the same period, who 'tried to show that Dürer's art theory had evolved from primary notions about rhythmic formality' (Davis 2017, 241). For Kaufmann, 'proportion and other systematic pictorialized constructions like linear perspective are merely epiphenomena of visual rhythm' (241).

It is automatism, however, that externalises and extends relatively non-conscious rhythms into conscious action. This is because automatist

Figure 3.4 Len Lye, *Untitled (Sea)* (c. 1930). Pencil on paper, 25.4 × 38.5 cm. Digital Image: Govett-Brewster Art Gallery © Len Lye Foundation Collection.

drawing or doodling involves holding writing, drawing or painting tools, allowing the hand to wander in its mark-making and trying to suppress or disperse imagistic sequiturs or logical inferences.

The artist Len Lye writes, 'When I'm in the mood to draw I cultivate a vacuous, seaweed-pod state of kelp in my skull. Attached to a pencil, I doodle in a bemused attitude' (quoted in Govett-Brewster Art Gallery 2018). Figure 3.4 is a drawing from 1930 of sea urchins or seaweed, which shares movement-percepts (identifiable motor imagery such as waving, flicking, swimming) similar to the style of Miró and his marine animals and toys of the same period. But Lye arranges the spindly drawings to appear to vibrate, and they resemble well-known scientific drawings of neurons and axons which were circulating at the time. Lye believed he was using the 'old brain', associated with strong emotions and instincts, to draw rather than being inhibited by the 'new brain' (prefrontal cortex) which supports following premeditated plans.[11] This kind

---

[11] Perhaps more than any artist Len Lye was obsessed with the modalities of rhythm across many media. I hope that the ideas presented in this book will inspire researchers to engage with his work in an interdisciplinary manner.

of automatism shows that mind wandering is not all in the head, that its rhythms can become externalised and made into objects or drawings and picked up by the viewer to stimulate their own mind wandering. Important for allowing this wandering to occur is the ability to feel or project movement in a static medium. Abstract details, non-human shapes and forms in clumps of rock and piles of dirt can engage motor imagery.

Paul Klee's often repeated comparison of drawing a line with going on a walk – 'An active line on a walk, moving freely without goal. A walk for a walk's sake' (Klee [1925] 1972, 16) – expresses the dispersal of rational will that is focused on analysing form. For him, one eye sees and the other feels. As a musician and artist, Klee rigorously pursued the usually tacit feelings viewers have that drawings and paintings possess rhythmic and musical qualities. Klee wrote that 'there are paths laid out in an art work for the eye to follow as it scans the ground rather like a grazing animal . . . . The image is created from movement, is itself fixed movement and is recorded in movement (eye muscle)' (quoted in Düchting 1997, 67). It seems important that following the line like a grazing animal is not about scanning the image for information about directions as we would with a map, animated by the concept of a purpose or destination. As Heidegger puts it, referring to Klee's late paintings, 'we should want to stand before them for a long while – and should abandon any claim that they be immediately intelligible' (Heidegger 1972, 1).

Indeed, it would be nonsensical to read Figure 3.5, Klee's *Twittering Machine* (1922), as if it were a map. The viewer follows the suggestion of certain slow rhythms associated with a sinuous line, interrupted by crossed lines and triangles. A crossed line can be read as a jolt; a triangle becomes lighter by tapering off as it rises up; several triangles and lines together are rhythmic. The viewer is familiar with the hand and wrist movements that might have been used to create certain knots or turns in the line, and draws upon muscle memory and motor imagery to imagine drawing the line themselves, which adds to the sense of rhythm and movement experienced in viewing the work. Accompanying this are iterated and arbitrary movements of the head and eyes, with attentional blinks or pupil dilations (what Klee refers to as 'eye muscle'). It is as if we are extensions of the twittering machine. Apart from a playful, rhythmic sensibility, Klee is able to convey more complex musical concepts. The twittering machine suggests both simultaneity and sequentiality.[12] The

---

[12] The interest in music and painting in this period in Europe is intense, as we can see from the example of Wassily Kandinsky. Kandinsky had a long-standing friendship with Schoenberg and was influenced by his rejection of melodic and thematic composition. Düchting mentions rather more geometrical treatments by Adolf Hoelzel, Johannes Itten and Richard Paul Lohse. Robert Delaunay's interest in simultaneity, bright colour

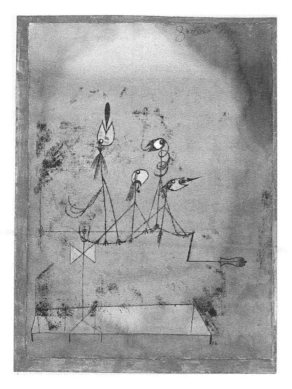

Figure 3.5 Paul Klee, *Twittering Machine* (1922). Watercolor and ink; oil transfer on paper with gouache and ink on border, 63.8 cm × 48.1 cm. Image: Museum of Modern Art, New York City/Scala, Florence. Out of copyright.

soft-focus background tones provide an unchanging key, a relaxing sense of inertia. The bluish-green tinge remains a dampening presence on the edge of consciousness, while the zigzagging lines and spikes of the machine stimulate feature detectors in rhythmic pulses of attention. Imagining the sounds the birds might make adds a further dimension to the feeling of things happening at the same time. One goes back and forth following the lines. The handle of the lever one is tempted to turn using muscle memory suggests the birds will alternate their positions of high and low, providing

---

[12] repetition and variation was shared by Klee. In his lectures at the Bauhaus, Klee explicitly dealt with the topic of rhythm. Klee began his lectures on this topic by 'demonstrating how several parallel lines combine to form simple patterns, which he termed "structural rhythms"' (Düchting 1997, 33). This 'analytical' form of thought is in contrast with Klee's famous statement 'drawing is taking the line for a walk'.

levity and further cues to think rhythmically. Interludes of restful 'grazing' alternate with thoughts about how the apparatus might 'work'. We use reason to understand how the machine is built in terms of push and pull, interpreting the lines as functional parts of the machine, yet this kind of thought tends to dissolve when we leave function behind to surrender to the washes of colour and the charming birds.

The resort to the Latin origins of abstraction, *abstractus*, to withdraw or draw away from, does not tell us what one is drawing away from. It could be a mental state that examines ordinary concrete objects (things and, by extension, images of these things) to reason how objects are given to us in experience, or it could involve following arbitrary associations in pipe-dreams while looking at the same objects. In other words, abstraction can be arrived at logically or non-logically, and certain kinds of art seem to emphasise these different tendencies towards abstraction. The geometrical art of Piet Mondrian is normally associated with the logical construction of abstraction, while the more spontaneous works of Wassily Kandinsky or Jackson Pollock seem to arrive at abstraction through improvisation. Complicating this logic/improvisation dualism is the claim that both routes lead to the absolute or the spiritual, the 'not of this world'. It could be said that European abstraction took a 'down-to-earth' route, emphasising matter and the oneiric, while American abstraction looked to metaphors of light, immateriality and purity.[13] The absolute can be understood in this context as nothing ever before conceived, the 'unprethinkable', the kind of imageless thought that mathematicians or logical positivists understand as propositions and equations, unsullied by images and metaphors. Abstraction has also been associated with reductionism: to abstract from something means to reduce or discard details for a general outline, simplifying aspects such as structure, rhythm or densities of light and dark. But this, paradoxically, would also mean paying attention to, rather than withdrawing from, the thing to be reduced in this manner. As Varnedoe writes:

In fact this distinction between reductive and productive ideas of abstraction has been a bugbear in the history of art. When the French abstract artists of the 1930s

[13] Of course, the 'spiritual' in Kandinsky and Mondrian, and in early European abstraction associated with esotericism and theosophy, emerged from historical conditions. Polcari uses the term 'spiritual' many times in his book on American abstraction. While some artists undoubtedly understood their art from this perspective, Polcari implies that it is possible to observe the spiritual in artworks as some kind of content, even though the works identified as such do not resemble each other in the least. 'Spiritual' can mean a great many different things. It can be associated with a surge of oceanic feeling in a picture filled with great detail, yet such an idea might also come to mind in contemplating a vast emptiness. For the trope of spirituality and its association with purity, hygiene and 'high art', see Poor (2007) and Cheetham (1991); see also Gooding (2001), which also often adopts the word 'spiritual' as a given, as a property or end produced by art.

tried to form a group, they got caught up in a huge debate about what to call their movement: they could only agree on a hyphenated term, '*abstraction-création*,' distinguishing between those who were distilling forms from visual experience and those who claimed that they were creating pure forms not derived from vision or nature (Varnedoe 2006, 47).[14]

The hyphenated form *abstraction-création* brings together reductive and productive ideas of abstraction. It seems to suggest that one can be creative in being reductive by drawing away from the observation of objects and things, nature and the visual world, or by paying careful attention to how they suggest rhythms that are experienced in the gut.

For Wilhelm Worringer in *Abstraction and Empathy*, abstraction arose as a result of 'man' turning away from the world rather than wanting to fully take part in it: 'Naive sensuous oneness with nature is replaced by a disunion, a relationship of fear between man and world, a scepticism toward the surface and appearance of things' (Worringer [1908] 1997, 102). He writes:

Aesthetic enjoyment is objectified self-enjoyment . . . to enjoy myself in a sensuous object diverse from myself, to empathise myself into it. What I empathise into it is quite generally life. And life is energy, inner working, striving and accomplishing. In a word, life is activity. But activity is that in which I experience an expenditure of energy (Worringer [1908] 1997, 5).

Perhaps, by extension, this kind of 'self-enjoyment' can also happen with objectless matter as a participation in its rhythmic complications, the way matter moves, and matter can move in ways that are agreeable or not. For Worringer, abstract art is a kind of withdrawal from the arbitrariness of phenomena, an extracting of certain regularities and rhythms in somatic experience consonant with order and repose. Abstraction rids form of its contingency and eternalises it. Worringer may have understood the expenditure of energy of the somatic constitution as somatosensory rhythm, muscular movement and memory. In abstract art, 'self-enjoyment' arises in a tendency of idealism but it is also a form of introspection and withdrawing from the world. Influenced by this idea, or perhaps by the darker musings of Schopenhauer, Klee wrote of abstraction:

The more horrible this world the more abstract our art, whereas a happy world brings forth an art of here and now. Today is a transition from yesterday. In the

---

[14] This is in contrast to how Varnedoe understands other kinds of abstract art as a language to be translated or decoded, or as a word game or puzzle, and even a system of self-references. He thus sees the play of irony and abstraction in Cy Twombly's, Jasper Johns's and Gerard Richter's 'art about abstract art': witty references to historical precedents that create an 'autonomous system' and a 'symbolic game' (Varnedoe 2006, 270–271) or a 'learned language' (41).

great pit of forms lie broken fragments to some of which we still cling. They provide abstraction with its material (Klee 1964, 313).

This is certainly a way to understand Kurt Schwitters's *merz* work (from '*kommerz*' or commerce, with associations of filthy lucre), which brings together the clash between high and low. Materials found on the ground, urban detritus, decaying matter and shrapnel were arranged into abstract designs, 'dirtying' art but also giving voice to the feeling that the terror and destruction of World War I had brought civilisation to its knees. In Schwitters's later works, the collages become darker and are dominated by dirt and soil. Dirt and besmirched ideals, remnants of people's lives and traces of bodily fluids come together in loose compositions that suggest the influence of gravity. We imagine looking down beneath us, as if we are standing on the painting, where the remains of civilisation are scattered on the ground as if after some violent event. In Schwitters's works we are tempted to piece together the shards of people's lives in nonlinear ways, joining together the abject and melancholic with the facticity of matter, and this was groundbreaking for matter painting.

3.6   *Informel*

Briony Fer claims that some abstract works are engaged with symbolic structures (Fer 1997, 120) and perhaps reflect the death drive (73) or complications of a castration complex (88, 104). The psychoanalytic approach takes matter and folds it into rational or semiotic analysis as an externalisation of latent tendencies. This approach is particularly adept at cutting short the sensations and rhythms of art by ruminating on the death drive or erotic drives or by analysing their visual equivalents in shapes and symbols. Krauss acknowledges the different ways in which the unconscious could be understood by psychology as opposed to psychoanalysis. The former 'veers off from psychoanalysis in that it posits no mechanism of repression ... It was psychoanalysis that would view the unconscious as divisive, as the turbulent source of a conflict with consciousness' (Krauss 1993, 137). Meanwhile, the 'conflicts' that perceptual psychology deals with are things like binocular rivalry and visual illusions such as the rabbit-duck diagram. Some of Krauss's interpretations of art blend aspects of psychology and psychoanalysis, but mainly she favours explicating structures of repression and the influence of underlying drives that deposit clues in conscious image making. The problem with this approach is the appeal to conventional conceptual personae – the Oedipal complex, the pleasure principle, the death instinct and so on – which are easier to identify in a work that appears to contain symbols or figurative elements than in those abstract paintings that appear to depict a void or striations in a rock. To view abstract works in this way would involve plumbing the depths of matter painting to find only pareidolia or the hidden clues of a Rorschach test.

More convincing is the idea that matter painting involved 'painters who were actually showing their broken spirits through the dismembering of modernist tradition' (Guilbaut 2007, 47). The issue of whether abstract art is to be understood as a withdrawal from the world into intellectual introspection, or as a direct or indirect expression of the world's brutal facts, has been discussed by Richard Storr. For him, works composed of burned straw, mud and dirt, trash and urban detritus

vividly recall the radically transformative impact that the worst of times left on those touched, if not permanently marked, by the seizures of the twentieth century, and they are a testament to the regenerative capacity of the imagination after it has stared into the abyss . . . . That the work itself is almost entirely abstract dictates that interpretation steers clear of explicit storytelling or semiotic reductivism. Yet the fact remains that in the minds of its initial audience as well as of its creators, recent exposure to organized violence compounded by the impending threat of the still greater violence of a hypothetical but not at all improbable World War II haunted the surface of these works to an extent that exegesis based on innovations in pictorial conventions and process alone cannot satisfy the alert reader or viewer (Storr 2012, 256–257).

Dubuffet, Burri, Fautrier, Wols and Tàpies are some of the prominent artists associated with matter painting, Art Brut, Tachism and *Informel*, although Fautrier and Burri rejected associations with the term *Informel*.[15] These artists did not form a coherent group with a manifesto but they shared common approaches in their various practices. Heterogeneous materials associated with the ground, the earth, sand and dirt were incorporated into their paintings at a time when the purity and aura of oil paint were still paramount. These admixtures consistently created cognitive dissonance for the viewer unable to resolve the

---

[15] There were other artists associated with *Informel*. Dutch-born Belgian artist Bram Bogart's work became well known for its heavy impasto, bulging with extravagant colours and textures, which seem to be applied with a piping nozzle of the kind used for icing a cake. The works from the 1950s, such as *L'Univers Relevé* (1959), are more sombre and intricate palimpsests of lace and viscous foam, frost and ejaculate. In the 1960s the artist arranged his canvases on the floor to allow gravity to play on the oozing thick paint, poured from above. Another Dutch artist Jaap Wagemaker's early works are in keeping with the Italian artist Roberto Burri's in using cloth, twisted and dipped in paint, to suggest old clothes, cord or scum washed ashore from a toxic lake or caught up in a swamp. His later works are more deliberately structural and cluster together a series of rings and lenses that suggest futuristic technology, eyes, microbes or aerial views of factory chimneys. The decay and sedimentation of complex structures seem to be a strange attractor in all these works.

conundrum that the art was also earth. The artwork, framing a cultural world, was traditionally understood as an articulation of the specific historical conditions prevalent at the time of its production. If art was understood as the articulation of a world, dirt was the denial of this articulation. Yet dirt actually *was* an accurate reflection of post-war Europe, reduced to rubble, and these matter paintings appeared to be taking a sample of the reality around them to become an extension of the debris. Reducing art, and by extension culture, to its material substrate seemed the most powerful way for these artists to express their speechlessness, following Jung's suggestion that 'upheaval in our world and the upheaval in consciousness is one and the same' (quoted in Polcari 1991, 113).[16]

This complements Storr's suggestion that the value of European postwar abstraction has something to do with its enigma, which

resides in a manifest materialism addressed to a crisis of faith in transcendence of the kinds that had hitherto been the primary object of nonobjective painting. The work was objective in the extreme, if by that one understands objectivity as an insistence on the priority of the phenomenological attributes of formats, surfaces, and structures, their tactility over and above their opticality ... their way of operating as portals to spaces beyond our ken even as they concentrate the viewer on what is directly within reach, and finally their abidingly strange anti-illusionism in contrast to their strenuous evocation of horizons that exist only in the imagination (Storr 2012, 256).

But perhaps this non-dualistic thinking that takes abstraction and the concrete as aspects of the same reality is not a recent phenomenon after all. In the earliest cave paintings, the stone surfaces on which they were made became the ground or filling for the figures; lines were drawn using pigments from the earth, with the rough, irregular shapes of the surfaces showing through the contours of bodies. Ancient pottery, and clay pigments for frescoes, also allowed unadorned areas of earth, sand or soil to remain visible. These no doubt provided encounters that could remind viewers of the ground under their feet and the humble origins of all human activity. These ideas were pursued with much enthusiasm in Dubuffet's Art Brut, which resembled cave paintings, described by him as an attempt to 'rehabilitate the dirt' (Foster et al. 2016, 337).

The ambiguities between matter and form, figure and ground, were nowhere more enchanting than in the caves. As Gerald Cupchik remarks,

---

[16] This is also captured by the artist Roberto Matta, concerning his painting process: 'I thought at one point that the introspection transformed itself into world introspection' (quoted in Polcari 1991, 28).

assuming that these prehistoric caves were places where torches were lit, and given the fractal and hypnagogic effects of a flickering fire, it is not difficult to imagine the oneiric power that these rhythmic surfaces and shadows wielded, rhythms given new life in the viewer's imagination of muscle movement and gesture (Cupchik 2016, 282). The temptation is to overstate intentional and willed shape-making. Cupchik agrees when he proposes that cave paintings were not the result of 'acting mechanically to approximate conventionally fixed images' (274). Instead, what is enabled in these performative and sensorimotor situations is involuntary thought and the ability to daydream with eyes wide open, using the ambiguities of the cave's surfaces. He also suggests that, in the firelight, 'private images of the wandering mind have an opportunity to gain expression' (282). This dialogue between mind and stone was animated by the fire. For Bachelard, the human mind, poetry and knowledge were all developed in meditation before a fire. This could perhaps be explained as a way of learning to sustain the tension between the mindfulness of observing the shadows and effects of the fire and dissolving into mind wandering: '[T]he dreaming man seated before his fireplace is the man concerned with inner depths, a man in the process of development ... the flame comes forth from the heart of burning branches' (Bachelard [1938] 1987, 56).

3.7   André Masson

The work of André Masson in the 1920s also contained elements that were later to be found in matter painting. Masson was physically and psychologically scarred by World War I, having served at the Somme. After a long period of rehabilitation, his academic style of painting moved towards the looser improvised metamorphic forms of Surrealism. Masson expressed the fundamental themes of sex and death, using a rare combination of sand and dirt and automatism. But this automatism was still haunted by traditional, cultural memories of figure drawing. Short, erratic rhythms were suggested by dotted lines, ticks, angular diagonal lozenge shapes and zigzags. In *Figure* (1927), a striking series of jagged or vibrating lines in blue and red suggests electric currents or energy signatures in the joining together of figures, with overlapping zones filled with smooth sand. The artist was drawn to images of fire. He adopted a densely patterned, rhythmic, calligraphic style where shapes and forms broke off into tall, branching flames and complex clouds of smoke, often suggesting erotic motor imagery, twists and thrusts, arms encircling waists and thighs. Masson's erotic tongues of fire were described by Sylvester as 'staccato flicks and jabs which make the erogenous zones electrically alive' (Sylvester 1997, 452). This clustering together of energies, images, sensations and hand movements exercising memory and the imagination was entirely in keeping with Bachelard's suggestion that 'every struggle against the sexual impulses must then be symbolized by a struggle against fire' (Bachelard [1938] 1987, 102). This is because the celebration of passion in Masson's works is never far from destruction. Masson does not represent light or fire, as such, but burning. Yet one has to become attuned to the complexity of rhythm that keeps us looking into the fire in order to paint it, to imagine the eyes burning. Death was often shown as the history of events, mythologies, battles, birds hunted by predators, animals devouring each other, depicted with forms that resembled shrapnel and torn limbs. He often developed forceful juxtapositions of images, the kind of contrasts found in Bataille (whose work he illustrated), using the irregular rhythms of automatic drawing.

William Rubin identifies in Masson an obsession with 'transmutation' in these years. Discussing a sculpture with the title *Metamorphosis* (1927), Rubin describes it as 'a curious small plaster of an animal-mineral-vegetable being in the process of devouring itself, intended as an image of "perpetual metamorphosis"' (Rubin and Lanchner 1976, 126). But it is in his marine imagery, slipping sand onto the canvas, that Masson achieved not only a metastability of these forces but also a remarkable originality that influenced later artists, particularly in bringing together frank and self-evident matter (sand, the ground, stone) and form (snaking lines that seem to emerge from the dunes). The metamorphosis of forms is sometimes also associated with Heraclitus, and indeed Masson executed a painting with that title in 1939.

In many of Masson's automatist drawings, twisted lines carry the variable force of the artist's resolve, concentrated in tight areas, or flailing helplessly in areas where he has abandoned control, relying only on instincts but nevertheless informed by the practised routines and gestures of figure drawing. The sand tells the tale of a different flux, the textured surface of the grains, rough and smooth, greater and lesser than human intervention, speaking of deep time, the time of minerals and sedimentary layers, and the incessant waves. The sand as ground and the line as the crest of a wave alternate dispersal and focus respectively. Masson attempted to empty the mind with sand and Surrealist automatism, and with the psychology of Zen, to free it from rational constructs. He said of his automatism: 'Make a void in yourself, primary condition, according to the Chinese aesthetic, of the act of painting. If that seems inaccessible to us, it can at least serve as the point of departure of a wandering in space' (Ades 1994, 24).

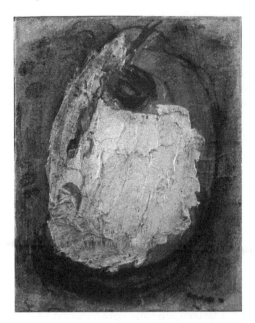

Figure 3.6 Jean Fautrier, *Otage n°7* (1944). Oil on paper mounted on canvas, 35 × 27 cm. Digital Image © Private Collection/Scala, Florence © Jean Fautrier/ADAGP. Copyright Agency, 2020. (A black and white version of this figure will appear in some formats. For the colour version, please refer to the plate section.)

3.8 Jean Fautrier

In his 1940s *Otages* (hostages) series, Jean Fautrier created images in response to disturbing recollections of the cries of victims tortured by the Gestapo in a forest outside the psychiatric hospital where he took refuge during World War II. The images are created with a thick paste, shaped with a palette knife to form heads, animated here and there with minimal distorted features, sometimes a twisted mouth or a darkened eye. Dead matter is able to express the artist's memories of pain, terror, dying words. We gain a number of insights from Curtis L. Carter's description of the techniques adopted by the artist. Fautrier began in the following manner:

Gluing rag paper (often the discarded pages of books illustrated by Fautrier) to canvas with an adhesive made from fish-skin scrapings. Fautrier applied a thick white primer to the surface and made a preliminary drawing with a light oil glaze. He covered this initial drawing with a layer of *enduit*, a coating used to prepare walls, to form the *haute pâte* structure. He worked this material with a palette knife, spoon, or trowel until he achieved the right surface, with a frostinglike texture that was firm enough to the artist's marks and gestures, yet supple, allowing maximum freedom in the execution of the painting. At this point, Fautrier again drew and re-covered the motif with *enduit*, which he might do repeatedly. He often manipulated the surface with the wrong end of his paint-brush, giving the pastel tracing a paintlike quality. In this final stage, he sprinkled crushed crystals or powder in various shades across the worked paste surface (Carter 2002, 21).

This description reveals a process that shares some similarities with traditional papermaking, relief plasterwork for walls, fresco painting and even aspects of cookery (sprinkling, icing). With their rough centres, and lines painted across them, the works also remind one of cave paintings and graffiti, arbitrary stains and plaster repairs on ancient walls. For Alexander Iolas, the *Otages* evoke the 'color of weather-stained masonry, of mineral deposits left by water seeping over stone' (quoted in Carter 2002, 21). The twisted plaster filigree in many of Fautrier's works, particularly *Trapeze* (1958), reveals an interest in pareidolia, medieval marginalia and grotesquerie, as well as the Surrealist traditions of decalcomania and metamorphosis. The muddying of categories of activity suggested by such a heterogeneous combination of processes is accompanied in the viewer's mind by a complex set of categorical disruptions. The critic Jean Paulhan wrote, '[T]here is something injuri-ous (and I would say almost filthy) in Fautrier's canvases' (Paulhan 2002, 53). The viewer experiences a 'shaking and oscillation [that] warn us of the ambiguity of a painting atrocious and gentle at the same time, whose colors change meaning before our eyes' (53). The percep-tual multistability in many of Fautrier's works arises from the tension of matter/form, dispersed vision (global)/focus (local) and face/landscape, as well as bottom-up/top-down stresses. On the level of aesthetics, multistability is manifest in the troubling encounter between atrocity and beauty, which the material processes of these works help to sustain as cognitive dissonance.

It is interesting that for Karen Butler, interpreting Paulhan, the physical properties of the matter help to create an experience for the viewer, a 'moment of disclosure', which evokes 'feelings of pleasure and repulsion ... disturbing his ability to think rationally. It is precipitated by the material aspects of the painting. This is what Paulhan calls an event (*événment*)' (Butler 2002, 47). Butler suggests that Paulhan tries to capture the irrational sensations the work produces by using an enunciative or performative description of sensations and movements to rekindle their affective rhythms and patterns. She mentions his description of the paste as 'masticated, ground, tamped down', which, in French, can be alliterated using hard 'gr', 'br' and 'cr' sounds, and which we associate with grinding, scraping and scratching. The image and its poetic description both release a rhythmic, pulsing, affective experience across the different senses: aural ('listening' to or imagining the words spoken about the image in Paulhan), visual (the image itself), haptic (the consistency and viscosity of the material) and motor (the reading of gesture and rhythm). This is taken even further to the olfactory domain, comparing Fautrier's work with cats that follow a 'ritual way of covering up their excrement with some earth, some cinders (then they sniff)' (Ponge 2002, 49).[17] This provocation, which is also circular and rhythmic, introduces other appalling aspects to Fautrier's affective painting: the smell of excrement, fish, a rotting corpse, creating a strong impression of the abject that the American critics of *Informel* found objectionable. And yet this kind of work expands our knowledge of what is possible with abstraction. Fautrier's various manual processes of feeling and manipulating plaster and paint, glue and powder, figure and ground, line and plane, are modelled by rhythmic finger movements and pressures. Rather than decoding them as semiotics, a key point to acknowledge is how the viewer instinctively experiences many of these complex rhythms.

Storr writes that one cannot simply look at Fautrier's paintings 'as precursors to the purely formal evolution of abstraction toward gesture, process, and inchoate matter. In them, the *informe* not only plays an art historical role, it also speaks of historical savagery and suffering and lives out its bitter fate' (Storr 2012, 252). The reference to Bataille's *informe* is telling because the way in which Fautrier has allowed the facial form to remain relatively uncomposed produces an ill-defined unease. The working with the hands is an interaction between the application and relaxing of the will, one that deposits a certain complexion in the matter that we detect as nervous tics, contortions of the face, down there, on the ground. Benjamin Buchloh, putting right an American tradition of misunderstanding Fautrier, identifies an intention 'to *yield* to the automatism of matter rather than to perform an automatism of graphic gesture' (Buchloh 2002, 66, my italics) and he also writes of 'ruins of gestures', which I understand as habitual actions we return to or choose to abort.

In Fautrier's work the attractor ruin, the gesture, is an integral part of the face and yet it is also the artist's hand touching and creating the face. The ambiguity is enough to activate areas of the fusiform gyrus and gamma frequencies, which help to fix and ascertain facial or object

---

[17] Although presumably they sniff and smell nothing at all if their actions have been successful, and this somehow is analogous with sniffing a painting.

percepts. Meanwhile, Van Leeuwen's work on perceptual multistability would suggest that it is possible the viewer's default mode network shifts the dynamics to broader associations (Van Leeuwen 2007). This is poetically described by Paulhan, for whom the reverie produced by these paintings contains 'sparkles of precious minerals, mosaics and illuminated manuscripts, ceramics and jewellery, soot-covered lacquer ware, a furnace and phosphorous. You think about it, but not without some uneasiness' (Paulhan 2002, 180).

Many writers and artists in this period were interested in mineral imagery of this kind. Bachelard quotes from a great number of them in both of his earth books. He describes a dream narrative by Michel Leiris that allows us to snag the rhythms of the oneiric which Fautrier captures, and this is to do with a figure or a body becoming-stone, becoming-mineral:

I had to move in an element that was increasingly viscous: I was not swimming in a river but rather in the earth, swimming between its stratified layers. What I had taken to be foam was but bubbles of crystal, and the seaweed suffocating me was in fact fossil fern imprinted in a coal seam. . . . In order to make my way forward, my hands have to thrust minerals aside that were incalculably deep; I slipped along amidst gold-bearing sands and my legs were covered in clay (quoted in Bachelard [1948] 2011, 170).

Such sentiments approach a kind of daydreaming state. As Paulhan writes, Fautrier's paintings are things we might imagine seeing when we

wake up in the middle of the night or too early in the morning . . . a zigzag, a flash, a burst of light, something like a hedge of twigs, bits of squares and lozenges (which come from the wainscoting on the wall), a cloud from which luminous rays stream? What? Debris, waste (Paulhan 2002, 186).

The overall image can be viewed in terms of the viscosity and tactility of paste and the way in which the configural view (and coarser segmentation) encourages a release from fixating on facial details. This point is particularly pertinent when viewing the works of Frank Auerbach. My own viewing process reaches saturation when, switching rapidly between face and matter, these distinctions appear to dissolve into a general state of abstraction. The ethical frame retrieves the subject of physical torture. The facial recognition structures sabotaged by the artist's guile persist, even though the face becomes unrecognisable when horribly contorted, 'the mutilated, stiff, bloody, truncated body; the crushed, deformed, swollen, tumescent head. Moral torture at the same time, as if it were a question of the just or the innocent – it is the expression of these bodies and these faces' (Ponge 2002, 173). As with so many artworks, perceptual multistability is also loaded with moral connotations. The rhythmic struggle, the failure to retrieve the face, is passed down through history from artist to viewer via the material interface.[18]

Buchloh provides an astute description concerning the detachment of the linear 'structure' (drawing) from the ground. The drawing is supposed to contain the matter, colour or content of the object, the *haute pâte* (paste), but in fact fails to do so; the impression is often that the line wanders away from the matter or even defaces the matter as a form of graffiti. This is an intuitive and artistic translation of the philosophical problem of matter and form; the first of these terms Buchloh associates with the somatic and corporeal, the latter with the semantic and abstract, creating an ambivalence. A similar ambivalence is pointed out by Paulhan. This metastability continues with other phrases which draw on the particular historical context, such as 'facture and fracture'. Buchloh concludes that the *Otages* paintings

are not portraits, but they are pictures of the dead; they are not figurative delinea-tions but they are automatist depositions of matter and texture; they are not paintings that depict, but neither are they self-reflexive modernist abstractions. This structure of the neither/nor, the either/or, is what we would call now – with some necessary opacity – the circularity of Fautrier's negation (Buchloh 2002, 69).

This circularity continually denies a fixed state of cognition, a conclusion or resolution. It is dynamic and can be understood as a feedback loop. Let us remind ourselves of the insight that circular causality is able to offer. Lewis's approach in emotion research stresses that

a coherent, higher-order form or function causes a particular pattern of coupling among lower-order elements, while this pattern simultaneously causes the higher-order form. Cognitive scientists have begun to model higher-order mental states such as intentionality and consciousness as emergent forms that constrain the activation of the psychological or neural constituents producing them (Lewis 2005, 174).

In other words, rather than just seeing the neural circuits as causing higher-order states, one should think of higher-order states, and access to the world outside, as constraints on the patterns of neural activity. What Buchloh intuits is a recurrent negation of familiar concepts to do with how we should understand abstraction through formalism or depiction in terms of verisim-ilitude. The work is somewhat figurative and somewhat abstract; it is matter painting in the sense that matter seems to be hovering in between these two

---

[18] It will not be lost on philosophers that this begins to sound like a comment on Immanuel Levinas and his work on how the encounter with the face of the other involves a whole set of ethical challenges.

technical art historical terms. In fact the suggestion is that this particularly cognitive, top-down process of applying technical terms is in continual flux or upheaval. Fautrier's work sustains or refreshes the metastability between the different levels. At the 'lower levels', neurological activity, oscillatory frequencies and neurochemical processes are already metastable, ready to switch to moments of stability or to continue with further instability. This metastability affects and is sensitive to perceptual routines, such as motor imagery, which feed back rhythms, intervals and variations to the lower levels. There are adjustments to the body, such as different kinds of eye movements, pupil dilation, pulse and breathing. These adjustments will interact with the lower levels, and with higher levels in terms of feelings and emotions that have their own uncertainties and cycles, and these can influence the retrieval and formation of concepts. Importantly, mind wandering and reverie are manifest on a special 'level' of activity, equally engaged with the multi-level causality described, yet heterogeneous in the sense that it can recruit various higher-level states for a higher-level metastability. There will be a complex switching between and smearing of emotions (with Fautrier, unease, sadness, disgust, reverence), memories and concepts (autobiographical or semantic, war, cultural memory, abstraction versus figurative, matter and mind), the oneiric and the nonsensical (rhythmic, imagistic, relatively formless or fractured). If one attends to matter painting it will provide some constraints on mind wandering. Go too far into ordering this experience with critical analysis and one will be distracted from participating in a certain dynamic heterogeneity. For then we will be drawn to a system of rules and dialectics governing conceptual relations – a world of geometry that matter painting does not possess.

Bachelard supposes that with certain reveries 'it seems that every element seeks either marriage or struggle, episodes that either calm or excite it' (Bachelard [1942] 1999, 13), to create the consistency of instability: 'paste' (*la pâte*), which Bachelard identifies with the very notion of matter found in Dubuffet and Fautrier. Paste is where the formal and material imagination come together, where sculpture is the result of matter 'both resisting and yielding ... it amasses all ambivalences ... I cannot emphasize too much how important the experience of fluidity and pliability is to an understanding of the psychology of the creative unconscious' (13). For Bachelard, 'there is no reverie without ambivalence, no ambivalence without reverie' (13). This is more than just a witticism; it reveals that rational control, the kind I have described throughout this book, is about making sense, matching unfolding experience to prior knowledge and conceptual systems in order to do so. In turning towards this we simultaneously turn away from what Paulan, Buchloh and others have tried to express: the power of association and affect.

If 'affect' is an intense, formless yet embodied process that lies under, distorts or even escapes control, it is reinforced by paintings that are produced by the same process. The hybrid qualities of texture, rhythm or granularity are left to run their course, and they resist judgement that would inhibit their development in time and complexity. When cognition attempts to make sense of affect, or to capture it, it gives it a precise structure, a boundary and a name: it becomes a particular emotion such as grief, sadness or joy. A painting might also aim to provide easy recognition of emotions using a set of precise structures. Fautrier manages to suggest that this precision is failing. In this view, painting is a struggle, a physical one between emotion and affect, capturing and releasing. As Roland Barthes writes, 'Each of us has his own rhythm of suffering' (Barthes 2010, 162), and each of us will choose a different medium through which this rhythm finds its measure.

Deleuze and Guattari write:

A child in the dark, gripped with fear, comforts himself by singing under his breath. He walks and halts to his song. Lost, he takes shelter, or orients himself with his little song as best he can. The song is like a rough sketch of a calming and stabilizing, calm and stable, center in the heart of chaos . . . . A refrain is a territorial assemblage. It marks off territory in chaos (Deleuze and Guattari 1988, 311).

If Fautrier's work is a struggle between emotion and affect, form and the formless, then this struggle is a territory occupied by art or, as Heidegger suggests, art is where the struggle between the world of form and the formlessness of matter is staged.

There is another fundamental refrain in Fautrier that the paintings about death are very much also about life, the artist's own rhythmic sensations, actions and patterns of energy co-extensive with the matter of painting. This is not to be understood simply as a signature or gesture, but as an acceptance of choppy matter to intervene in history as it complexifies or undermines our cognitive assurances about art.

The cutting, smearing, caressing and twisting hand movements with the palette knife are discernible and can be associated with methods of torture that we can only imagine. The rhythms of trying to escape from the strange attraction, and returning to it, are the rhythms we see in the execution of the works, in the manual and tactile engagement with the material. These fluctuations seem to propagate at the level of material, sensation, percept and concept, with memory triggering repeats and after-effects. The non-linear dynamics are complex and are not limited to one image alone. Leaving one painting to go onto another picture in the series creates an attractor ruin, a resonance assembled again for another hostage: the sense of a series, one hostage lined up after another, adds to the gloom.

3.9   Jean Dubuffet

The artist Jean Dubuffet's exclamations about art are vigorous:

Far from being filled with wonder at human reason, I find it one of the poorest and dullest things there is . . . . Man does not amaze me in the least in this world where I am amazed by so many things! Trees, for example, trees amaze me and fill me with wonder. Man does not at all. Basalt, ah yes! Basalt stupefies me, I fall on my knees before basalt! (Dubuffet et al. 2006, 140)

In the 1950s, taking his lead from Fautrier, Dubuffet's early works adopted quantities of earth that appear as a ground or wall. The flat-plane, stone-face tableaus appear to be scratched upon by a child, mad-man or caveman to resemble graffiti. In his later work, mark-making, considered irrelevant and unworthy of being looked at, dissolved into quantities of earth, soil, dirt and sand that seemed to make their own marks. We see this with the *Matériologies* and the *Texturologies* series, developing his *haute pâte* technique:

mixed from tar, asphalt and white lead, occasionally enriched with cement, plaster, lacquer, glue, and even lime, sand gravel ... applied to various supports like chipboard or canvas. Spoons, knives or palette-knives replace the paintbrush to shape the substance and to work on *modulating* the surface before it hardens (Messensee 2003, 34, my italics).

'Modulating' is a key descriptive term for the technique: an even spreading of granular texture without favouring any particular area, shape or form that would suggest a focal point. In this even-handed way, the artist's soothing circular action is similar to plastering, sanding or spraying, which smooths out the calligraphic. This modulation of material creates an optical flow that allows the mind to drift.

Of particular note in this period are *Texturology I* (1957), with delicate dried grasses and snowy effects comparable to Tobey's micro-aggregates, and Figure 3.7, *Soul of the Underground (L'ame des sous-sols)* (1959), predominantly browns and greys with silver foil, resulting in a fine sheen with sparkling areas to suggest schist in cave, lit up by a torch. The striations, pockmarks and erosion in these works, the marks of deep time without an author, can also be read as delicate and rhythmic traces of the artist's fingering and pressing of inanimate matter.

In Figure 3.8, *Place for Awakenings* (1960), Dubuffet aimed to stimulate the viewer's eye to pan over the surface of the work, rhythmic movements one might associate with hypnagogic states of delta oscillation, punctuated by attention-seeking details that could trigger gamma frequencies. As Dubuffet suggested, 'the work of art directly captures the operations of

Figure 3.7 Jean Dubuffet, *Soul of the Underground (L'ame des sous-sols)* (1959). Oil on aluminium foil on board, 149.6 × 195 cm. Digital Image © Mary Sisler Bequest, Museum of Modern Art, New York, USA/ Bridgeman Images © Jean Dubuffet/ADAGP. Copyright Agency, 2020. (A black and white version of this figure will appear in some formats. For the colour version, please refer to the plate section.)

the mind, in the same way that the electrocardiogram directly transcribes heartbeats' (Dubuffet in Loreau 1965, n.p.).[19]

We might also remind ourselves of the non-logical possibilities of the work as a product of the artist's access to oneiric rhythms and patterns. Perhaps this kind of content, which I intuit in my encounter with these works, could best be expressed in poetry. For example, Rainer Maria Rilke writes:

> Perhaps, through difficult mountains I make my way
> In hard veins and, like ore, alone;
> I am so deep that I see no end
> and no distance: all becomes nearness
> and all nearness becomes stone   (quoted in Bachelard [1948] 2011, 171).

There is a sense in which the figure and ground in Dubuffet's *Texturologies* are so entangled that all nearness becomes stone; we feel entombed, as if we were deep inside the earth. Our eyes need to adjust to the subtle dark tones, and the adjustment augurs a feeling of sinking and enclosure. This journey

[19] From a letter to H. Damisch, 7 June 1962.

Figure 3.8 Jean Dubuffet, *Place for Awakenings* (1960). Pebbles, sand, and plastic paste on board, 88.4 × 115.2 cm. Digital Image © The Museum of Modern Art, NY. Gift of the artist in honor of Mr. and Mrs. Ralph F. Colin/Scala, Florence © Jean Dubuffet/ADAGP. Copyright Agency, 2020. (A black and white version of this figure will appear in some formats. For the colour version, please refer to the plate section.)

of the imagination becomes even more compelling as it seems to have a connection with presentiments of oneiric depths and rhythms. Reveries and daydreams of matter, stone and rock, mineral and earth, modelled in sculpture or in painting, share mental rhythms and textures that are also expressed in literature: they 'take root in the deepest layers of the unconscious' (Bachelard [1942] 1999, 4), leading us to dream not just of their outer appearance but of their depth and texture. And 'it is when we dream of this interiority that we dream of the repose of being, of a deep-rooted repose, a repose that has intensity, that is not merely the totally external immobility prevalent among inert things' (4). The imagination 'can give rise to endless oscillations, oscillations that penetrate the minutest detail of the interiority of substances' (33). According to Bachelard, 'when we dream depth, we dream our own depth' (38). If this is true, then what happens when we observe such images in art? Are we engaged in outward visual perception while at the same time going deeper into introspection?

For his *Texturologies*, Dubuffet employed the so-called Tyrolean technique, used by plasterers and stonemasons to disperse surface details.

The technique consisted in:

shaking a brush over the painting spread out on the floor, covering it with a spray of tiny droplets. This is the technique, known as 'Tyrolean,' that masons use in plastering walls to obtain certain mellowing effects. But, instead of brushes, they use little branches of trees – juniper, box, etc. – and they have different ways of shaking them to get the particular effect they want. I combined this technique with others – successive layers, application of sheets of paper, scattering sand over the painting, scratching it with the tines of a fork. In this way I produced finely worked sheets that gave the impression of teeming matter, alive and sparkling, which I could use to represent a piece of ground, but which could also evoke all kinds of indeterminate textures, and even galaxies and nebulae (Selz and Dubuffet 1962, 137).

To disperse or scatter pigments helps to initiate a restless circulation of mental images that arise without any particular order. Tiny sparkling quartz or silicon particles or brick dust seem to free-fall into visions on a grand scale – the great outdoors, alluvial valleys, cosmic nebulae. The technique aims to achieve an 'indeterminate texture' but this kind of reverie on different scales can be interrupted by a focus on the details of facture, cogitations about how an effect is achieved. Perhaps there is in this kind of abstraction more of an uneasiness, a struggle between attention paid to material facticity and poetic associations that complexify this attention. In this sense, the twofold action of attending to something while turning away from something else becomes a manifold complexity. Dubuffet states: 'Art should emerge from the material and the tools, and carry traces of the tool and its struggle with the material. The human being should speak, and so should the material' (Messensee 2003, 34). And yet, while considering the details of 'the material', one is also involved with forms of abstraction:

An internal mechanism should be set in motion in the viewer so he will scrape where the painter scraped, scumble, gouge, fill in and bear down where the painter did. He will feel all the painter's gestures repeated in himself . . . . The viewer will experience the viscous pull of gravity where the paint has run and when brilliant outbursts occur he will burst with them. He will dry out, contract, fold into himself where the surface is dry and wrinkled. And he will swell up suddenly inside at the sight of a blister or some other sign of death (Dubuffet, quoted in Solomon R. Guggenheim Museum 1981, 8).

In *Texturologies*, Dubuffet suggests that viewers reach

a dangerous point at which the object is liable to function like a thinking machine, like a screen of meditations and visions, and the most lowly object devoid of all interest. It is easy to understand that people who are enamoured of art become alarmed when it is taken to such an extremity that the distinction between that which is art and that which is nothing at all threatens to become uncomfortable (Dubuffet and Husslein-Arco 2003, 120).

Whereas abstraction is commonly understood as a process of turning away from the concrete, here the details function as anchors for concepts while sensations are being felt.[20] And when we are forming concepts, the sensations can be returned to by visiting the same features in the painting that triggered the sensations. Through the variations in matter that the artist has preserved or altered, the artwork will help orchestrate a subtle interplay of concepts and sensations, what we associate with mind wandering. Sensations seem to fade more slowly than the lightning flashes of thoughts, so there are some interesting overlaps that could be extended in abstract art.

The lithographic prints of Dubuffet's *Phénomènes* series (Figure 3.9) were images of soil, mould or erosion, fine and coarse textures taken directly from the ground, the walls and from gravel on the gallery floor, and transferred onto the lithographic stone plate. Thus the works have a ready-made aspect to them. A total of 362 prints were produced. Many of them feature an all-over modulation of earthy, stony stochastic distributions. One is left with the artist's strange attraction, perhaps shared by the viewer, to an abstraction that is characterised as wordless thought. As Dubuffet puts it, 'art should be born from materials and should borrow its language from it. Each material has its own language so there is no need to make it serve a language' (quoted in *The Times*, 2008). Dubuffet's suggestion here is not about a language of reason but about letting matter speak to us in *its* own language, insofar as we feel its rhythmic textures and contingencies that splinter language as we know it.

With the *Phénomènes*, the eyes become accustomed to the gloominess, and when a percept does seem to emerge it brings to an abrupt end a cycle of mind wandering and the pleasure derived from the abstraction of continually dissolving activity, which seems self-organising, 'of its own accord' – or at least our projection of its autonomy is at the root of the pleasure we derive. The most reduced perception we are able to distil: a flat picture, a ground of dirt, variations of colour and ink momentarily slacken the great imaginative power of the mind's abstraction. Yet it is by looking down at the desolate ground and accepting its meaninglessness that the processes of abstraction begin. The facticity of the flattened soil exerts a kind of lethargy that is hard to remain with for long. One can only look at grains of sand for a certain amount of time, empty of images. The mind wanders from this zero-point as a kind of reaching towards new

---

[20] Psychologist Edwin Hutchins writes that 'a mental space is blended with a material structure that is sufficiently immutable to hold the conceptual relationships fixed while other operations are performed' (Hutchins 2005, 1562). Thus features of abstract art could act like markers for concepts, or the relations between them, that need to be kept in place or returned to while the mind wanders or shifts attention to sensations.

Figure 3.9 Jean Dubuffet, Plate from *Les Phénomènes* (1958–1959). Image: Private Collection/Christie's Images/Bridgeman Images © Jean Dubuffet/ADAGP. Copyright Agency, 2020.

life, without even being aware of the chaotic itinerancy from thought to thought that ensues, because after all the image we are looking at has not changed or moved. The artwork is revealed as it is: unchanging matter before us. We can concentrate on this thought alone, but it would require suppressing the patterns and associations we have discovered, those we are discovering and what we are about to discover.

If we were to close our eyes and try to imagine how an abstract concept such as metastability could possibly appear to us in image form, we could do no better than to take an example from the *Phénomènes*. There is a continual de-differentiation, produced by shifting aggregates or tendencies of detail, which is simply the switching of different kinds of perceptual grouping, general or particular. This alone is enough to act as a kind of metronome for the appearance and disappearance of the earth.

3.10   Antoni Tàpies

Antoni Tàpies was said to have

embedded the humble materials of the street – sand, dirt, clay, rock, pebbles – into his paint. Their surfaces resemble walls, windows, and doors inscribed with graffiti-like letters, numbers, crosses, and other signs, pockmarked with bullet holes, and worn with abrasions. Indeed, scratched, pinned, and gouged, these wall-like surfaces conjure memories of the damaged, discarded, and abandoned architecture of his beloved Catalonia (Schimmel 2012, 197).

It is not insignificant that Tàpies was fond of quoting Heidegger's concept that 'the origin of the work of art is art' (Tàpies et al. 1988, 58). Tàpies pursues ideas in his practice that complement Heideggerian philosophy. In 'Painting and Void', where Tàpies discusses the stillness of Japanese art, he quotes Heidegger: '[E]xistence is the extreme nothingness which is simultaneously copiousness' (Stiles and Selz 1996, 58). Anthony Lack offers a Heideggerian framework for engaging with Tàpies's work:

The tension between the rich earthiness of the color and background material and the lowly junk produces a distancing, an alienation effect that causes one to step back and consider our civilization and its garbage. We look at the works and we see an intermingling of beauty and destruction, of delicacy and waste. We see our modern relationship between earth and world and it may give us pause and stimulate thinking about other types of relationships (Lack 2014, 78).

But perhaps this interpretation is itself cognitive and willing – *Vorhandenheit* – rather than surrendering to a will-less state of being with the work, particularly in the case of Tàpies's 'wall' paintings, which support such a 'way of being'.

In his writings, Tàpies recalls that in Zen Buddhism the Bodhidharma was renowned for his meditation, facing a wall for nine years (Tàpies [1970] 203, 117). Decades after producing his wall works, in 'Communications on the Wall' the artist describes the inspiration that helped to produce them: 'I was absorbing everything from archaeological treatises to the counsels of Leonardo da Vinci [presumably his writings about stains on walls], from the destructive spirit of Dada to the photos of Brassaï' (Tàpies in Ishaghpour 2006, 115). The artist goes on to explain how his graffiti/wall works captured a busy painterly technique of 'frenetic movement [and] gesticulation' when the 'unending dynamism of those gashes, blows, scars, divisions and subdivisions that I inflicted on every millimetre, on every hundredth of a millimetre . . . suddenly took a qualitative leap' (115). These feverous efforts are characterised as an existential struggle:

One day I attempted to reach silence directly, with greater resignation, surrendering to the fate that governs any profound struggle. My millions of furious clawings

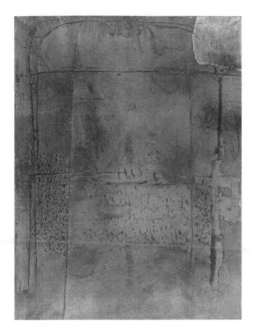

Figure 3.10 Antoni Tàpies, *Ochre gris* (1958). Oil paint, epoxy resin and marble dust on canvas, 260.3 × 194.3 cm. Digital Image: Tate Modern © Fundació Antoni Tàpies, Barcelona/VEGAP. Copyright Agency, 2020. (A black and white version of this figure will appear in some formats. For the colour version, please refer to the plate section.)

became millions of grains of dust, of sand . . . . A new geography lit my way, carrying me from surprise to surprise: suggestions of unusual combinations and molecular structures, of atomic phenomena, of the world of galaxies or of images in a microscope . . . of ashes, of the earth from whence we come and to which we return, of the solidarity born when we realise that the differences among ourselves are like those between one grain of sand and the next (Tàpies [1970] 2013, 20).

This remarkable passage suggests an almost mystical vision that begins with a struggle with the work as a present-at-hand encounter that is exhausted by countless 'furious clawings'. This is also a physical surrender where distinctions between artist, practice and matter, 'differences among ourselves' and between grains of sand, all become a clearing of marks and distinctions for *Dasein* to emerge. And this way of being is by no means fleeting. In a 2006 interview, Tàpies confesses that 'with time I've come to see the unity of all things ... that the cosmos is a mass of matter struggling with itself, positive and negative' (quoted in Lack 2014, 75). Here, the underlying drive for some artists is revealed: it is not just aesthetic but also ontological, with a different understanding of matter, closer to its natural state, as science reveals to us.

Providing some clarity on how this is so, Umberto Eco sees abstract art not as the expression of new scientific concepts but as 'the negation of old assumptions' (Eco 1989, 90). Matter paintings function as epistemological metaphors that

represent the repercussion, within formative activity, of certain ideas acquired from contemporary scientific methodologies – the confirmation, in art, of the categories of indeterminacy and statistical distribution that guide the interpretation of natural facts. Informal art calls into question the principle of causality, bivalent logics, univocal relationships, and the principle of contradiction. This is not the opinion of a philosopher who is determined to find a conceptual message in every form of art. This is part of the poetics of the artists themselves, whose very vocabulary betrays the cultural influences against which they are reacting (Eco 1989, 87).

The artistic re-evaluation of humble matter as something worthy of contemplation is a transformation in artistic practice as much as it is an act of cultivating humility. It is perhaps some kind of intuition of a non-cognitive and chiasmic being-as-matter and matter-as-being. The feeling is more easily and directly experienced by returning to the artworks, which provide a clearing for this to happen. Tàpies was also fond of an aphorism by Chinese artist Shih T'ao: 'I speak with my hand, you listen with your eyes' (Ishaghpour 2006, 145).[21]

Meanwhile, Heidegger suggests, '[L]et me give a little hint on how to listen. The point is not to listen to a series of propositions, but rather to follow the movement of showing' (Heidegger 1972, 2). The movement of showing involves setting into motion iterative and varied saccades, and because the artwork has no beginning, middle or end there are infinite variations that are possible in observing the work. I have already mentioned how, in the *Open Work*, Eco compares this kind of matter painting to a road which is extremely open and contains a maximum amount of information:

We are free to connect the dots with as many lines as we please without feeling compelled to follow any particular direction . . . . The eye no longer receives any direction . . . . This is again evidence that the richest form of communication – richest because most open – requires a delicate balance permitting the merest order within the maximum disorder. This balance marks the limit between the undifferentiated realm of utter potential and a field of possibilities (Eco 1989, 98).

And, Eco explains, the 'reader' is excited by

the new freedom of the work, by its infinite potential for proliferation, by its inner wealth and the unconscious projections that it inspires. The canvas

---

[21] Tàpies, quoting Shih T'ao in an interview with Manuel Borja-Villel.

itself invites him to avoid causal connections and temptations of univocality and to commit himself to an exchange rich in unforeseeable discoveries (Eco 1989, 91).

Tàpies's works appear as a simultaneity of possibilities. At its highest reaches, abstraction claims its own reality, sui generis, rather than merely acting as a sign that services a conceptual reduction.[22] This is because 'concepts, by their very nature, ignore details that images incorporate' (Bachelard [1948] 2002, 200). It is also because the details of Tàpies's abstraction act as phenomenal triggers, not just because of the concepts to which they might refer but also because of the juxtapositions of textures and tears which are felt as rhythmic and tactile, providing concepts with the background hum of sensation. This may be what Heidegger means by will-less knowing, when 'thinking is no longer a representing' – a knowing that does not try to force its will on objects, 'staying open' to the possibility that they 'no longer have the character of objects . . . . Not only do they no longer stand counter to us, they no longer stand at all' (Heidegger 2010, 74–75). All of these material variations sustain a phenomenal richness, and conceptual and cultural associations, but without recourse to words arranged on a page in sequential order. Importantly, these experiences, which Heidegger would want to pursue as examples of being rather than examples of consciousness, occur with alternation of willed and will-less engagement.

3.11   Alberto Burri

This movement of showing can be understood as disclosing the matter in art, which nevertheless remains withdrawn or one step ahead. What is revealed in the Heideggerian sense is this withdrawing from the grasp to follow the lead of matter's movements, which seem to express our own. This dance is a fugacious entity that both human matter and non-human matter seem to share, for a while at least. After World War II, Burri went on a visit to Paris to see Fautrier and Dubuffet's *hautes pâtes* and developed a way of exploring and preserving in his work a greater range of materials such as polymer, plastic and plaster. *Composizione* (1953), consisting of burlap, thread, polymer and gold leaf, appears to be a canvas erased of any imagery or composition in paint. Yet, while presenting this raw matter to us, it nevertheless produces imagery and composition.

---

[22] The philosopher Theodor Adorno explains this with much cogency:

> Although the demarcation line between art and the empirical must not be effaced . . . artworks nevertheless have a life sui generis. This life is not just their external fate . . . . But the emphasis on the artifactual element in art concerns less the fact that it is manufactured than its own inner constitution (Adorno 1997, 5–6).

The Trauma of Painting (2015), a study of Burri's work based on a major retrospective at the Guggenheim in New York, suggests that a key technical and psychological approach in his practice involves a 'whole chain of pulls and tensions', as the artist himself puts it, and a 'precarious equilibrium and masterful control of the unpredictable' (Braun et al. 2015, 36). Burri's practice is carefully poised between premeditated construction of formal composition and bricolage. The term 'precarious equilibrium' is useful to describe the kind of abstraction where the clear distinction between intentional and non-intentional aspects of the work is undermined. The stitching and patching together seem intentional, revealing manual and perhaps fallible attempts to repair cloth, but fabric itself is perfectly and mechanically produced. Yet even here, there is something interpersonal and somatic about the threadcount. In moments of shared intimacy, an embrace perhaps, sometimes even the tiny pattern of thread running through a jacket or shirt is the only thing we need to return us to the maze of memory.

As a doctor in the Italian army during World War II, Burri witnessed the suffering of wounded and dying soldiers directly and on a daily basis. In 1943, he became a prisoner of war in Hereford, Texas. Being an officer, Burri was not forced to work and found a way of dealing with his time and place in the world through painting, sculpting and collage. This eventually became so important to him that he abandoned medicine and practised as an artist for the rest of his life. Burri's '*sacchi*' (sackcloth) work in the late 1940s and 1950s consisted of a series of burlap patches stitched together and sometimes stained with drops of red pigment. The sackcloth must have occurred to Burri as a material that not only had multiple conceptual associations and strong emotional registers but was also eminently practical and logical – canvases too are a type of cloth, albeit of a finer weave. Sometimes the cloth is retrieved from sacks of American foodstuffs given to prisoners of war in the rubble of the cities of Europe under the Marshall Plan, but mostly the sacks were saved for him by friends from his home town in Italy. The convenient, down-to-earth material was also historically specific, used in makeshift hospitals as partitions. Importantly, the sackcloth quite readily stimulates haptic sensation and evokes associations with patting or stroking cloth in a gesture of sympathy or tender care, a cloth that may conceivably cover skin. Burri's use of stitching, particularly the subtle ways in which he left the stitches visible, could refer to the rags of the countless poor and destitute in the streets of post-war Italy, but to Italian eyes the tears and sackcloth were also symbols of the humility of Franciscan monks in early modern Italian painting.

Although Burri was trained to use a scalpel and suture wounds with dexterity, and even though many of his *sacchi* have been understood as a reference to this skill, there are two distinct psychologies that the stitching suggests: careful and intentional stitches emphasising neat borders and curves, employing a precise technique to achieve a controlled outcome, versus more erratic, even careless tacking. The first psychology leans on linear perceptual surveillance aligned to motor control, while the second arises from the relaxation of the first, towards a high tolerance of 'crude' zigzagging workmanship. What seems consonant with the two different psychologies ensnared in this kind of abstraction is the idea that healing requires patience and 'letting be', acceptance of time and discomfort, whereas its opposite involves activity associated with rumination: picking at a scab. Through the alternation of precise stitching and relaxed 'spontaneous' tacking, which preserves open areas where paint is left to congeal, Burri succeeds in sustaining an unsettling interplay between binding and unravelling, controlled activity and the relinquishment of control, repairing tears or

Figure 3.11 Alberto Burri, *Composizione* (1953). Burlap, thread, synthetic polymer paint, gold leaf, and PVA on black fabric, 86 × 100.4 cm. Digital Image © Peggy Guggenheim Collection, Vercelli/ Scala, Florence © Fondazione Palazzo Albizzini Collezione Burri, Città di Castello. Copyright Agency, 2020. (A black and white version of this figure will appear in some formats. For the colour version, please refer to the plate section.)

In the *sacchi* we also read the semiotics of the world of textiles and sewing, the different thread counts and weaves of the sackcloth chosen for the works. The artist incorporates fibres and stained rags that already have their own patterning inscribed on them before being added to the artwork. One patch could be coarse-grained fabric intended for foodstuffs, another could be fine-grained twill that was once part of a pillowcase, sometimes with traces of hand-sewn embroidery. Each patch of cloth has a different tempo of facture, complicated by the artist's own erratic running stitch, cross stitch, back stitch, tacking and tearing, bringing together different groups of rhythms. Eye movements, motor imagery and haptic engagement work with pulses, smoothly continuous and circular, the contiguity of the cloth disrupted by folds, stains or tears, its continuity broken into sequentiality. The viewer sees a particle or a wave, a point or a relationship. The 'diacritics' or 'accents' of the breaks, stitches and tears provide focal points that arrest the eye, while the larger expanses of cloth appear to dampen activity, allowing the eye to gloss over the details in waves of sensation.

In Figure 3.11, *Composizione* (1953), the zigzagging thread is sometimes 'violent' and sometimes 'delicate', according to Braun et al. (2015, 160), or stitched 'tenderly' or 'furiously'. Studies by Helge Lundholm (1921) show that individuals draw rounded and curved lines when asked to express adjectives such as 'gentle' or 'lazy', and zigzagging lines in response to 'furious' or 'harsh'. Similarly, Shigeko Takahashi (1995) finds that art students drawing abstract lines are understood by the majority of observers to be communicating concepts such as 'anger', 'human energy', 'joy' and 'tranquillity'. These studies suggest that the *sacchi* works can be understood using emotion words that have rhythmic connotations: slow, methodical, uniform and regular stitches that convey care and continuity, and loose multi-directional tacking that appears erratic, hurried, panicky, discontinuous. The result is a switching between two major rhythms, optical flow and optical breaks, providing momentum. Sometimes the stitching, just as with musical accents, can act as a boundary for groupings of rhythms. On this view, each grouping would be a particular resolution of granularity of the cloth: fine linen bordering the coarser texture of sackcloth, with both areas joined by the rhythmic staccato of a certain type of stitch.

In terms of shape recognition, some square patches seem resolute and solid, relatively unblemished, while other patches with curved edges seem to sag, or seem soiled or bloodied, stained by what appear to be body fluids, burned and twisted with pubic hair (Braun et al. 2015, 160). In some works the stitching is arabesque and knotted like a root or a calligraphic flourish. The 'movement' from which we derive rhythm is governed by the grids, repeated forms and uniform stitching, which suggest continuity, as much as it is by the ruptures of tears or drops of red paint which break up this continuity to create immediacy. This alternation of smooth continuum and isolated 'facts' can break restful and contemplative alpha frequencies with beta or gamma frequencies associated with attentive viewing.[23] These different ways of modulating rhythm and counter-rhythm occur through colour and textural effects (rough, smooth, matt, gloss), accompanied by brief glimpses of strangled twine or unhealthy stains. 'Mindful' inspection may just be a conscious interval in between a sensuous panning over the aggregates of detail considered in their simultaneity of co-locations.

Artists are usually acutely aware of how external matter and biological processes internal to the body (such as motor, haptic and rhythmic sensations) seem to arise out of a coordination of external and internal events. But sometimes these sensations are accompanied by concepts that seem to direct or alter their flow. Braun et al. suggest that Burri may have had some prior knowledge of how internal and external resources could be aligned through Enrico Prampolini's concept of 'arte polimaterica', which was said to arrange 'the transubstantiation of humble and ephemeral substances into abstract compositions of rhythmic and spatial harmonies' (Braun et al. 2015, 53). Braun et al. also cite the Futurist Filippo Tommaso Marinetti and his understanding of 'tactilism' as a way to engage with art, a kind of 'blindfolded "journey of the hands" over textured panels ... surface variables, vibrations, temperatures ("thermic sensations") of different materials' (53). With Burri's works, haptic response arises 'offline', through the imagination and through optical inspection. One is not allowed to explore these works by touching them, so the eye wanders as if touching them, although it can move to different areas much faster than a hand physically stroking the material. Even though phenomenally, in our mind's eye, we feel as if we are touching the painting 'long hand', we inevitably apply 'shorthand' skips and edits. But the artwork itself is also a series of cuts within fields of continuous granularity.

These variations in feeling rhythm and producing rhythm with matter in art occur without any exertion of the will. They seem to arrive of their own accord and in nonlinear fashion, iterated or complicated. This is not just 'pure' rhythm understood as alternation, sequentiality and simultaneity, but something that could be translated as the rhythm of thoughts arriving, departing and overlapping, which one is able to discern because of the visual anchors provided by the artwork. At the higher-order level of

---

[23] As Van Leeuwen and Smit report, large-scale patterns of alpha synchronisation tend to disappear with the interference of focused tasks (Van Leeuwen and Smit 2012, 138).

concepts, Burri seems to have achieved what every artist dreams of achieving: communication without the spoken word, to be understood in silence with the eyes and with the senses, which obey no sequential rules of unfolding. Burri's artworks can stimulate general, sensory and rhythmic explorations of different types of matter, separate from the kind of analytical thinking that classifies this matter with reference to cultural associations related to linen, sackcloth or plastic, because in his works these materials are freed from their usual role of supplying substance and texture to recognisable objects. This, in a nutshell, is another insight into what we might mean by abstraction. Empirical studies of how the brain processes images of materials (fur, ceramic, glass, metal) indicate that in the ventral stream, particularly around the fusiform gyrus associated with facial recognition, information about materials is constructed that is 'supramodal' (Hiramatsu et al. 2011). Supramodal information is general and abstract, not differentiated into experiences associated with a particular sense. This suggests that the kind of abstraction we see in *Informel*, Dubuffet's dirt paintings or Burri's *sacchi* stimulates general categorical information over and above the automatically grasped 'tool use' information implemented by the dorsal stream.[24]

In 1952, Burri created one quite exceptional monochrome, *Bianco* (now in the Collezione Prada, Milan), which he returned to in later years. It appears to show a cluster of craters, the rim of each circle frayed and crumbling, the intervals in between scored with many intersecting lines of craquelure. Executed around the same time as Dubuffet's *Phénomènes*, this work shrugs off painterly conventions and the formal aspects that Arnheim would have admired. It appears simply as a natural surface pockmarked by time and the forces of nature. Burri's long-standing interest in the rhythmic and textural possibilities of flaking and craquelure continued in later years with his *bianchi* works. These consist of a paste of industrial plaster-like material (vinavil) left to dry and separate into intricate networks of fissures and hairline cracks that fracture the image and remind us that it is matter. In Burri's 'material abstraction', the craquelure is divested of its percept (identifiable figure, form or 'scene', such as a face or an object) so that it appears as it is – a material that is cracked – but is also abstract, providing a formal complexity. In foregrounding this craquelure, Burri appears to let the matter in the work form its own shapes and their relations to each other. The rich patterns and surfaces that

---

[24] Areas in the dorsal stream do not contain information that distinguishes material categories. Obviously, in the experiencing of ordinary objects, these streams will cooperate seamlessly.

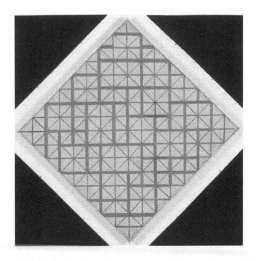

Figure 3.12 Piet Mondrian, *Composition in Black and Grey* (1919). Oil on canvas, 60 × 60.2 cm. Image: The Louise and Walter Arensberg Collection, 1950, Philadelphia Museum of Art, Pennsylvania, PA, USA/ Bridgeman Images. Out of copyright.

Burri's works preserve do not simply 'represent' fissures and broken layers; they provide an opportunity to rhythmically extend experience and being as part of seeing and perceiving.

According to Van Leeuwen (2007), with perceptual multistability we switch from the dynamics of one kind of pattern identification to another, and mind wandering allows the switching to occur. Although in reading exercises gaze duration and dwell times are affected by mind wandering (Schooler et al. 2011), there appear to be no studies of how saccades or pupil dilation change during mind-wandering episodes while looking at artworks.

From a purely introspective phenomenological perspective, viewing a famous work by Mondrian, *Composition in Black and Grey* (1919), Figure 3.12, one can become sensitised to how the painting causes mental rotation, pattern switching, and aggregation or disaggregation, and this switching feels rhythmic, a pulsating sensation that seems to occur without our control.[25] This also works with peripheral vision: if the viewer

---

[25] By introspective phenomenology I mean an intuitive technique of self-observation that can be cultivated while looking at art. Joined to a knowledge of well-known perceptual processes such as binocular rivalry, fine- and coarse-grained focus, colour perception and tactile response, this has also been called 'neurophenomenology', which 'stresses the usefulness of obtaining detailed, first-person reports of moment-to-moment subjective

fixates on one area of the image, a cross, for example, what happens in the periphery of eyesight? Pulsating patterns change; squares and rectangles overlap and seem to diffract. Here, Mondrian reveals 'an expansive compositional principle and a dynamic internal rhythm that is both optical and conceptual' (Crowther 2012, 32). Looking at serial repeats of crosses or a grid structure, can easily be intuited as a pulsing sensation even before it becomes explicitly conceptualised as such. It is interesting that this pulsing 'feels' like it is happening in the eyes.

The concentration of details in one area, alternating with sparser areas, can be experienced as a 'contraction' and 'release'. This is a rhythmic switching, as in Ehrenzweig's processes of differentiation and de-differentiation, and Van Leeuwen's understanding of the nonlinear dynamic switches that occur in perceptual multistability (Van Leeuwen 1998). We see examples of Op Art where wavy forms and repeated lines are easily and involuntarily felt as pulses and rhythms. Again, this is not just a metaphor. Kim and Blake (2007) show that the areas of the brain responsible for processing motion are stimulated by examples of Op Art. It is important that art seems to offer opportunities for becoming aware of and reporting on the embodied and phenomenal effects and anticipations that arise in the encounter with particular artworks.[26] Fundamental to all the senses is pattern recognition (we have seen how this is metastable in the olfactory sense with Freeman's studies, discussed in Part II). As Kulezic-Wilson writes, what is fundamental are 'the principles of creating rhythms of tension and release, alternating sections of density and sparseness and the rhythm of formal expect-ations and anticipations' (Kulezic-Wilson 2015, 67).[27] On the one hand, abstraction suggests turning attention away from the literal embodied experience and sensations triggered by the artwork, yet on the other hand, abstraction also allows us to ponder the complex rhythms that seem to cooperate between body and art.

---

[25] experience in order to uncover information about brain rhythms and dynamic coordin-ation relevant to mental functions and consciousness' (Fazelpour and Thompson 2015, 223).

[26] Again, the neuroscience supports this inference. Along with Varela, Thompson and others, Lutz et al. (2002) have long held that an important methodology for understand-ing cognition is to dig deeper into the relationship between behavioural, neurophysio-logical and first-person data. For them, 'the brain's response and its phenomenological correlate must result from the intertwining of the endogenous activity with its corres-ponding phenomenological distinctions and the peripheral afferent activity evoked by the stimulation' (Lutz et al. 2002, 1591).

[27] While the author emphasises the positivist perspective of gestalt psychology concerning making order out of chaotic stimuli with pattern-making, the artists of *Informel* were interested also in the opposite tendency: the falling apart of patterns into complexities that are at the same time less and more than pattern.

Figure 3.13 Simon Ingram, *Automata Painting* No. 3 (2004). Acrylic and gesso on canvas, 90 × 90 cm. © Courtesy the artist and The Fletcher Trust Collection. (A black and white version of this figure will appear in some formats. For the colour version, please refer to the plate section.)

It is interesting to contrast the early twentieth-century modernist sensitivity of Mondrian with a contemporary art practice, almost a century later, Simon Ingram's *Automata Painting* No. 3, 2004, because it shows how artists have developed even more complex understandings of abstraction and materiality. The title of the painting refers to cellular automata, modelling squares in a grid that produces a certain patterning based on some kind of simple rule, for example, whether or not certain numbers of neighbouring cells are occupied. Cellular automata produce a global complexity supervenient on local rules, and this principle has had an impact on modelling in biology (e.g. in the patterns of some seashells), psychology (crowd behaviour), particle physics and computation. Cellular automata have also been used to simulate neurons, where a neuron is considered to have a number of limited states which affect neighbouring neurons. In this sense, much like fractals (which cellular automata can produce), cellular automata act as a principle transferable to many different domains. Ingram's work transfers and adapts cellular automata to art-making.

What Ingram gives us is a glimpse of matter as both simple, in its principle of physical extension, and complexifying through time. At the philosophical level, the simple units and rules (in the artist's technique and in the viewer's viewing) are proto-phenomenal properties which produce phenomenal experiences of complexity. This phenomenal complexity is metastable; it appears as an involuntary shimmering or wriggling pattern. The shimmering 'in the eye' can act as a visual accompaniment to the concept, held in the working memory, of the micro-macro interplay between the simple level and complex emergent patterns. In the other direction, the configural, macro view of order disintegrates into focal points at the local level. This is analogous with Ehrenzweig's interplay between de-differentiating vision and differentiating vision, which the artwork helps to stimulate in its own manner. For Ehrenzweig this interplay is a kind of feedback loop, where 'outward perception and inner phantasy become indistinguishable, (Ehrenzweig [1967] 1993, 272). This indistinguishability is where nature, the artwork and the artist's consciousness disappear in an ongoing complexification of matter.

The issue is whether we side with the Kantian view, that our ways of modelling reality are modes of presentation ('for us'), or whether matter, the material universe, is organised along the principles of cellular automata ('for itself' – humans absented from the universe). It is a non-dualism that allows us to move beyond this quandary, if we understand matter in terms of 'neutral monism': the metaphysical notion that ultimate reality is all of one kind, neither mental nor physical, but neutral between the two. In the sense of art history, Ingram's work represents a complex behaviour emerging from the simple rules in Mondrian's practice, through 'agent-based modelling', where the modeller starts with micro-rules to explore the macro-behaviour that produces art. In this sense he is probably the first artist who has consciously shown us the moment in history when artistic practice migrates naturally into the fields of philosophy and physics to achieve, not only a complex interdisciplinary understanding of matter as cellular automata, but also the physical production of cellular automata matter in art.

Yet the *Automata* paintings are not seamless illustrations of a perfect system. They resist the Greenbergian modernism of pure opticality and, in fact, are not seamless at all – the edges leak, forming a messy edge that brings the universe of material contingency into the painting. This leaking is in excess of the unbearable perfection of the grid, it steps over the line, where the human initiates the act of freedom to disobey by ceding agency to the material. The artist allows the paint to function 'for itself', as a result of its own viscosity, interacting with gravity as much as possible. This suggests the work is non-correlationist in some respects. In the artist's own reckoning:

What is at stake here is a version of materialist philosophy being deployed in a pitched battle against idealism in painting. This idealism is of the kind that tends to separate thought from sensuous human activity, theory from practice, ideas from the domain of matter. In general, thought in the work of these painters occurs at the level of matter, it is matter. Thought is substantiated through folds, colour, flow, and occurs in this way and not abstractly. To call the artists' painting 'abstract' is in a sense to mislead, because this work is real or actual in a way that makes representational painting seem unreal (Ingram 2004, 35).

Other artworks that disclose the correlationist/non-correlationist rift in art are Burri's later *Cretti* works in the 1970s, which consisted of a series of white blocks of concrete-like material left to dry in the sun. Rent with fractures and cracks, the craquelure is quite distinct from manual mark-making associated with deliberately rendered surfaces. The network of cracks creates rhythmic complications and pulses, particularly when one intuits details in peripheral vision. The rhythms are jerky and irregular, yet sometimes seem to fall into interesting symmetries or repeats in a fractal manner. The mind wanders into various feelings and memories, adopting different scan paths, perhaps less attentive to particular shapes and more to a general panning over the surface when reverie begins. It is as if a desert is marked by drought and has been abandoned by all living creatures. The cracks seem to go their own way, a series of bifurcations quickly multiplying outwards and joining up with other cracks, a rapidly spreading self-organising – or self-fracturing – movement-matter.

Both Alice Aycock and Edward Weston were fascinated by the visual effects of cracked surfaces that seem irregular and arbitrary but also repetitive, rhythmic and textured. Landscape and 'abstract' photographers like Edward Weston, William A. Garnett and others worked in the traditional landscape genre with close-ups and cropping of alluvial riverbeds, dunes, twisted driftwood and the vistas of prehistoric glacial valleys, with Death Valley being one of the strange attractors for this photographic itinerancy. Along with chemical and mineral associations, these views of the matter-movement of rock and stone also support oneiric or moody reveries. Bachelard mentions Georgio Ascani's belief in the power of strata and earth to entrance us with 'opalescent fantasies' (Bachelard [1948] 2002, 143) and explains that, for Joris-Karl Huysmans, rocks that reveal 'metallic gangrene and petrified sores are more than picturesque excess; they express profound doubt about the integrity of all substances' (143). Bachelard makes connections with the underlying oneiric impulses of mind wandering with images: 'Chip away a rocky matrix, open up a geode, and immediately a crystalline world will be revealed to us; a well-polished section of a crystal shows us flowers, fine tracery, figures. Our dreams are endless here' (Bachelard [1948] 2002, 23).

We are dealing here with a kind of reality testing that puts matter 'out there' into a definite space outside our bodies. But we also follow the variable qualities of the materials, their breaking and branching, with rhythmic motor imagery that provides a certain pattern to our reverie. Thinking about geology and deep time, the patterns created by strata, activates a strongly rhythmic and abstract psychology where, as Smithson writes, 'the entire body is pulled into the cerebral sediment, where particles and fragments make themselves known as solid consciousness' (Smithson 1996, 100). The fascination here is with the multistability of matter's surfaces, from visions of illimitable scale to microscopic details, both uncountable, creating a multistability of scales. The image shifts from an intuition of the 'liquidity' and transparency of the camera lens, the liquid eye, to the dry and grainy photographic surface of the print, chapped skin, the cracked bones of the earth. The encounter is with objectless surfaces, used as a medium for the imagination to flip to dynamic conceptual transformations, settling upon them for a short while before switching to other possibilities. Each perceptual space triggered by an association (earth, cracked skin, valley, photograph) has its own set of rhythms, and the eyes adjust scan paths and resting points to accommodate each variation. The switching between associations is itself rhythmic. The imagination allows the image to be perceived at a particular scale, from prehistoric alluvial valley to pores of chapped skin to microscopic ridges of a photographic print. The images seem to present fractal possibilities, invariant patterns across differently imagined scales. Mind wandering helps the perceptual switching to occur, rather than fixating on one perception alone. Importantly, this kind of mind wandering does not turn attention away from the external image. Rather, introspection can continue and become more elaborate, and the image becomes the site or anchor for this wandering across points, aggregates, scales and textures. The complex perceptual multistabilities of the image sustain the eyes-wide-open daydreaming state described by Ehrenzweig, where 'outward directed perception and inner-directed introspection become indistinguishable' (Ehrenzweig [1967] 1993, 273).

3.12   Zen and Abstraction

As Curtis L. Carter (2002), Serge Guilbaut (2007) and Kent Minturn (2007) stress, the American critical reception of European abstraction in the post-war period was mired in nationalist ideologies of purity and modernity, influenced by the freshness of American art associated with bright colours, large-scale painting, ideas of painting's autonomy and the freedom of spontaneous gesture.[28] It is interesting that David Clarke, in his book *Water and Art* (2010), observes that it was mainly American Abstract Expressionism that focused on the imagination of water and liquids (Pollock, Frankenthaler, Louis, De Kooning), more than European abstraction. American modernism was held to be superior to European decadence, inward-looking cultural reflexivity, hesitation, miniature preciousness and dark 'regressive' colours. Greenberg was concerned with European art's reverence for good taste, while American art treated the canvas as 'an open field' rather than a 'given receptacle' extending the possibilities of the medium beyond traditional boundaries. American art was 'raw, fresh, spontaneous, and direct, as opposed to more conventionally finished French paintings' (Carter 2002, 23). For American critics of the time, such as Thomas B. Hess, Fautrier's work was mere confection, 'small, limited, for its textural precision' (quoted in Carter 2002, 23), while for John Ashbery, American abstraction had 'a swashbuckling energy [and] wide open spaces' (Carter 2002, 24). In its enthusiasm for instant gratification, bright open spaces, large formats and clear chromatic differentiation, this tradition of criticism was blind to how the eye becomes accustomed to the shade in order to discern smaller samples of matter, as we might imagine under a microscope, a new territory of the uncountable to be discovered. Perhaps the luminosity of American art must be wrested from its anti-value.

Mark Cheetham (2006) suggests we are still untangling the history of abstract art from certain modernist tropes to do with purity, autonomy and idealism. On this level of description, when we look at a painting, particularly the 'post-painterly' abstract works of Frank Stella or Joseph Albers, the painting can be seen as a pure image, regardless of its material facture. This may be the result of an automatic top-down generalisation or simplification of complexity to geometrical forms, drawing upon *a priori* schemata. Top-down processes can be automatic, but they are elaborated into a set of modernist values characterised by Greenberg's and Fried's approaches to abstraction as pure idea or transcendent absorption. A commitment to top-down processes would mean

---

[28] Greenberg wrote:

> The avant-garde arts have in the last fifty years achieved a purity and radical delimitation of their fields of activity for which there is no previous example in the history of culture. The arts lie safe now, each within its 'legitimate' boundaries .... Purity in art consists in the acceptance, willing acceptance, of the limitations of the medium of the specific art (Greenberg 2003, 557–558).

> For critical appraisals of this stance, see Marter (ed.) (2007), particularly the chapters by Guilbaut (2007) and Minturn (2007).

downplaying considerations of the bottom-up contingencies of fine-grained perceptual flux, such as the sheen on a canvas, the granularity of pigment or dust floating in the air of a room, all of which might be regarded as 'noise'. This point is well known in art history through John Cage's famous quip regarding Robert Rauschenberg's *White Paintings*: they are to be seen not as absolute white purity but as landing strips for dust particles. With bottom-up processes, we perceive all the fine-grained details as our eyes dart around erratically to discern abstraction in the fullness of information. With top-down emphasis, our sight comes to rest on a general vision of an ideal purity, light or darkness, empty of contingencies.

It is interesting that Harold Rosenberg's interpretation of abstraction as 'action painting' emphasises bodily movement as an expression of an existentialist struggle with matter. This is quite opposed to Greenberg's emphasis on calculation and intellectual detachment, which presupposes a disembodied eye, optical mastery and control over the body. This critical context in the 1960s formed a historically specific duality. Today we understand that cognitive categorisation and evaluation, the top-down control of Greenberg's approach, are actually cooperative with the bottom-up embodied approach favoured by Rosenberg, which valued the sensorimotor improvisation of the artist in response to the rhythms and textures of handling matter.

An important development that contributed to the emergence of abstract art was the influence of Zen Buddhist principles. After World War II, in haphazard and unexpected ways, Zen ideas spread to the United States and Europe with translations of Zen Buddhist works, publications of Zen paintings and visits from Japanese scholars. Artists read these books, went to talks and presentations at major colleges of learning, and attended Zen Buddhist centres, which became popular in the 1950s. In the world of music, the composer John Cage was a major figure interested in Zen, and he visited Black Mountain College where the scholar Daisetz T. Suzuki gave numerous lectures. For Cage, the non-dualism of fullness and emptiness is manifest in the paradox of sound and silence. A note held constant without variation can fade out of consciousness to become unnoticed and silent or, as witnessed in his piece *4'33"*, silence can be filled with the incidental noise that we usually suppress by concentrating on concepts. Many of Cage's works use chance procedures to subdue the mechanics of rational conceptualisation in order to compose/decompose sound in spatial, temporal and physical ways, liberating or foregrounding rhythm as matter – rhythm for rhythm's sake, as sound, not as meaning in a conventional narrative. Repetition is used to negate artful composition in favour of entraining the rhythms of breathing, the pulse or heartbeat, or to make associations with waves and walking. The 'emptiness' of music, its detachment from meaningful or expressive qualities, thus produces the fullness of embodied and natural rhythms in which music itself originates, providing pulsative duration and abstraction. In Cage, repetition is a fullness, and yet its lack of differentiation can be understood as emptiness. It is also a way to 'reset' a consciousness attuned to traditional modes of musical composition.

Helen Westgeest (1997) documents how the principle of 'emptiness as fullness' finds visual expression in the works of Mark Tobey, where countless marks, scribbles and rhythmic fragmented gestures suggest white noise. This kind of 'writing' offers comparisons with other artists such as Henri Michaux and André Masson with his automatic writing, a kind of 'nonlinear orthography'. The marks are bristling with life. As Tobey puts it: 'My work is obviously in a state of constant flux . . . I want vibration in it. Everything that exists, every human being, is a vibration' (quoted in Westgeest 1997, 75).[29]

Tobey's paintings are built up of accumulations of scribbling that suggest swarming on the fine-grained view or the kind of craquelure that caught Burri's and others' attention, unfolding into a dispersed global or configural view (Figure 3.14). The switching of focus back and forth creates a feeling of perpetual vibration. Perhaps the global view acts as an after-effect of the local focus or, in reverse, the local focus creates a delayed granularity in the global unfolding. This metastability is supported by the faster neural oscillations used in the binding of percepts, in tension with the tendency to disperse such percepts through slower frequencies that travel across broader areas of the brain. The rhythms of saccades, scanpaths and resting points are part of this broader rhythmic distribution. The movements of the eyes affect, and are affected by, hypnagogic states. Neurons, eyes, brushstrokes create a complicated set of interactions sustained over the domains of brain, body and world.

Zen Buddhism was founded by the Bodhidharma, who arrived in China in the seventh century. In the century that followed, its teachings centred on dissolving rumination by focusing on emptiness, and dissolving the self by accepting the higher principle of change. Zen Buddhism provided encouragement for some artists to explore matter and abstraction simultaneously. It provided an eloquent justification for thinking beyond dualisms such as mind/body, self/other, mind/matter. Many forms of meditation rely on the kind of alpha activity associated with calm contemplation and passive acceptance of things happening rather than the exercise of control and will power.[30] This feeds into an

---

[29] For recent research into the topic of Zen's influence on American art, see Levine (2017).
[30] As indicated in Part II and in Buzsáki (2006, 212).

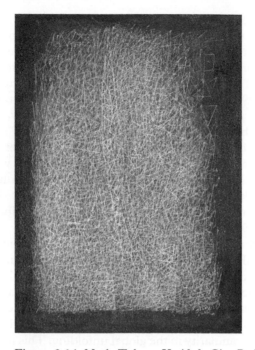

Figure 3.14 Mark Tobey, *Untitled, City Radiance* (1944). Gouache. 48.2 × 35.5 cm. Digital Image © Private Collection/Bridgeman Images © Mark Tobey/ProLitteris. Copyright Agency, 2020.

engagement with matter that is difficult, perhaps even unnecessary, to put into words.

Steven R. Pritzker writes that 'suspicion of language is a basic tenet of Buddhism' and it takes 'a great deal of training to calm what is called in India the "monkey mind," the constant chatter of thoughts' (Pritzker 1999, 539). In meditation the distinction between mind and matter, here and there, time and space, seems to dissolve or is dispersed in favour of a more fundamental reality.[31] The problem with trying to describe the unfolding of this nonlinear dynamic is obvious: explicit awareness of it, by which we might begin to put it into words, brings us out of it or codifies it into language and linear sequencing. Nevertheless, its general form has been described as reaching *mushin*, a state of 'no mind', and this may

---

[31] This could be expressed in another way. Gestalt theorist Albert Michotte maintains that our sensory experiences 'are infinitely richer in content than could ever be anticipated', and underlying these processes is 'a kind of "prefiguration" of abstract concepts, the mental "categories" of substance, reality and causality' (Michotte [1954] 1991, 44–45).

translate into the shedding of inhibitions and self-consciousness. Students of Zen Buddhism practise achieving *mushin* in painting, archery, calligraphy or swordsmanship, which allows them to become accomplished in these activities by dispersing rational self-conscious control in order to attain an embodied and automatic absorption in the use and acquired gestures of brush, bow, pen or sword. Becoming one with the tool, instrument or weapon was achieved by eliminating self/other, body/ tool and mind/matter dualisms that inhibit the flow of action. In European philosophy, we are familiar with this idea through Heidegger's notion of *Dasein*, a being absorbed in the world where distinctions between the body and pen, typewriter and tennis racket, dissolve in the moment of doing.

The aesthetics of *Sumi-e*, traditional Japanese ink painting, are based on the interaction between practice and spontaneity, a balance which places equal emphasis on the 'order' of form (objects, letters, things depicted) and the relative chaos of matter (the texture of the paper, brush, ink splashes, irregular areas, washes). Becoming attuned to this balance also requires switching from recognisable percepts (objects, letters) to relatively formless reverie (atmospheric washes, abstract shapes). To focus on particular details of an object requires the binding processes of faster brain oscillations, while reverie involves a slower bandwidth and less 'bounded' results.

Introspective phenomenology would describe viewing and thinking about a Zen painting as drifting, which one might experience in viewing peaceful mists and the rhythmic linear aspects of mountains, rivers and trees. The complication is that such drifting thoughts can occur in viewing the blend of techniques (wash, splash, rhythm, interval, texture) and the impressions that arise of basic shape recognition (rock, bird, mountain). The things portrayed as well as the way they are portrayed lead to drifting thought. Varying mood associations (excitement, relaxation, agitation) and sensations (softness, sharpness, smoothness, wetness) accompany and are dynamically intertwined with switches in attention to the different areas of the painting. One assumes that a painting arranged without too many breaks in rhythm would support a more stable mood. The switching and inter-digitation of these effects and investments in our encounter with the artwork are modulated by the artist's ability to combine and balance different techniques: a dark, linear, calligraphic rendition with a dry brush or point, along with lighter, broader and more evenly distributed washes, suggesting mist or the atmosphere.[32] A well-known example of this kind of painting is the work of

---

[32] It must be mentioned that a facility with the brush based on disciplined practice is already a kind of non-dualism. It takes years of study for an artist to be good at drawing and to acquire various kinds of painting techniques, while at the same time spontaneity and

master painter Sesshū Tōyō (1420–1506). Switching from specific (local) focus to general (global) focus, with each shift affecting eye movements, is a rhythmic transition the master artist is able to support and enhance through his own introspective phenomenology (mindfulness), which alternates with 'no-mind-ness' (*mushin*), psychological states that may accompany these kinds of techniques. This can be cultivated so that the mindfulness of the executive network does not interfere with the meditative insights associated with the resting state. Empirical studies have found that the kind of meditation that requires mindfulness, such as counting breath (as we see in forms of Buddhism), depends heavily on the executive system. This is opposed to the kind of meditation seen in Hinduism (Samadhi) and Zen Buddhism that favours the contemplation of nothingness and *mushin*, which depend on the default mode network (Tomasino et al. 2014).

It is interesting that Golding describes a Rothko painting as a 'gently pulsing whole' and quotes Rothko who wanted his paint to appear 'breathed onto the canvas' (Golding 2000, 218). According to the Zen approach, this breathing could involve the unity of introspection and extrospection. Daisetz T. Suzuki played an important role in introducing Zen to America. Many artists read Suzuki's books and, importantly for these artists, he described non-dualistic thought and being: the underlying continuity (or monism) of internal/external, mind/matter, subject/object, freedom/discipline, fullness/emptiness, one/many, stillness/dynamism – a non-dualism which can be discovered in viewing abstract painting. Interesting also is the fact that emptiness is associated with non-conceptual thought or 'emptiness' of narrative. 'Becoming one' with the painting process was also seen in various artists' practices. It is worth recalling Lee Krasner's comment about Pollock's painting that it breaks away from the concept 'that one sits and observes nature that is out there. Rather it claims a oneness' (Krasner in Belgrad 1998, 111).

At the root of this non-dualism is multistability, which works as a psychological state but is also exteriorised in the physical practice involved in executing the techniques of ink painting, in its composition and non-composition of matter and form, and in viewing such paintings where outward-directed perception and inner-directed introspection are blurred. In terms of empirical support, psychological studies of meditation show the adoption of 'a non-judgemental awareness of inner and outer experiences' (Chiesa 2009, 585) sustained by increased alpha/theta band activity associated with the default mode network. Embracing non-dualism is thus a physical activity at various levels of description. As the Japanese Gutai artist

---

[32] improvisation are highly regarded in Zen as well as in modern painting. The right balance between the two is a matter of metastability in execution.

Shiryu Morita explains, the psychological frame of mind for artistic execution is *'mu'* (emptiness), a 'positive negation implying "release" or "abandonment" ... a bringing to bear of the body of experience and letting-go-of-it' (quoted in Westgeest 1997, 206). The notion of letting go of composure in favour of flow, being-in-the-moment, open to chance and contingency, improvisation and implicit knowledge, also has affinities with Heidegger's *Gelassenheit* and *Dasein*, particularly the aspect he calls absorption through a ready-at-hand relation to things, which I would describe as being-in-matter or mind-wandering-in-matter. Along with Bergson, Suzuki explains that 'Zen is doing, becoming one with dynamism. . . . Life itself must be grasped in the midst of its flow; to stop it for examination and analysis is to kill it' (Suzuki 1994, 102). It is significant that the gesture of letting ink flow is a physical enactment of the releasement of will and the acceptance that the ink will, to some extent, find its own way. The non-dualism of practised releasement and the contingency of matter are the marks left on the paper. Helen Frankenthaler's work shares synergies with this psychological and technical artistic practice. The artist used acrylic or oil paint, thinned to the consistency of watercolour, on unprimed canvas and poured paint from cans, sometimes guiding the flow with sponges and cloths. Part of this artistic practice, picked up by the viewer, involves feelings about how matter has a consistency or moves in particular ways, which one anticipates and becomes attuned to in contact with materials. In such cases, art facilitates the releasement of matter to be itself, but this releasement deeply affects the art.

For Westgeest, concentrating on the fine-grained focus involved in painstaking acts of painting keeps artists' minds free from rational considerations and, by extension, viewers can also experience this benefit. Seen in this way, the activity is not far from some of the mystical exercises of repeating words or actions without linear rational rules in order to exhaust self-conscious inhibiting sequences and sustain nonlinear, spontaneous and involuntary sensations. In this way, abstract art helps the viewer avoid repressing spontaneous thought and opens up a fuller set of experiences. The painting and viewing processes both cultivate a kind of observation similar to meditation. Mindfulness can be used as a tool to disperse percepts, as it does in meditation when the aim is *shikantaza* (just sitting), just as counting sheep can be used to numb rational supervision or chatter. Agnes Martin writes that her painting was about formlessness, 'breaking down form. You wouldn't think of form by the ocean', and she wanted to encourage the eye to participate in 'free and easy wandering . . . a "holiday" state of mind' (quoted in Chave 1992, 145). In these different cases, visual experience involves forms of 'emptiness', or the non-dualism of fullness *and* emptiness, and in meditation this has been shown to lead to slower brain oscillations and theta activity (Chiesa 2009). This is comparable with Travis and Shear (2010), who describe 'automatic self-transcending', underpinned by the absence of focus, control and effort. This is said to involve 'letting go' of the experience and responding spontaneously with levels of passive awareness, often accompanied by feelings of well-being.[33]

In the relationship between meditation and art, which Zen encourages, the mind seeks equilibrium: too much chaos in the form of Tobey's or Michaux's crowded marks, or too much order as in Martin's grids, leads to a mind placed equidistant between order and chaos, mindfulness and mind wandering. In discussing the intricacies of mind wandering in Part II, I mentioned Mrazek et al. who suggest that such episodes involve a series of *partial* co-activations of mindfulness and mind wandering (Mrazek et al. 2014, 236). It is possible that mindfulness reduces the excessive chatter and rumination that are associated with mind wandering, allowing for more relaxed transitions. The way in which Tobey's and Martin's artworks are composed could facilitate transitions from mindfulness to mind wandering, and in this sense they could function as visual 'hinges' for the opening and closing of these states, comparable with the function of mandalas, *kasinas* or *yantras* in meditative practice.[34]

Another way of understanding metastability is not through the sensory overload of 'fullness' but with sensory deprivation. Ad Reinhardt emptied his canvases of all focused marks or gestures that could be read as being driven by ego or immediate emotions. His *Black Paintings* required significant resources of patience and discipline to produce, with multiple layers smoothed over again and again, and only slight variations in tone. It is with great effort that the artist achieves the appearance of nothing. While appearing to withdraw cues for percepts, gestures or rhythms – an emptying out of semiotics – the *Black Paintings* require the eyes to adjust slowly to the underlying geometric forms that seem to emerge from slumber. Just as with Rothko's canvases, Reinhardt's *Black Paintings* tend to undermine the conventional understanding of visual experience as grounding the relationship between observer and observed, suggesting unusual convergences of space, time and matter. These works call upon the viewer's own ability to exercise *shikantaza*. As Sylvester writes, if we 'risk a few instants of our busy time ... the matt black surface seems to sing' (Sylvester 1997, 69) – or, at

---

[33] The philosophical recognition of this is, again, Heidegger, for whom letting go is a kind of freedom from closing down possibilities with prior motives: 'Indwelling in releasement to the open-region would accordingly be the genuine essence of the spontaneity of thinking' (Heidegger 2010, 94).

[34] For an overview of this topic from a psychological and empirical perspective, see Schooler et al. (2014). One should remain critically aware of how mindfulness programmes support further integration into the rhythms of being 'productive', where the agenda is set by an institution, corporation or government.

least, to hum. This is a risk that becomes easier with practice; in releasing the work of art from its responsibility to signify, sacrificing time and meaning, one becomes enchanted. Sylvester goes on to write, 'I find it difficult to tear my attention away: my entire capacity for attention has been focussed into gazing at a contrast of colours that is scarcely perceptible and totally absorbing' (69). This multistability extends to the play of outward-directed perception and inner-directed introspection; the canvases can be read as breathable, immaterial darkness, easing the passage from internal to external and back again. These works require a stillness to discover their secrets: in time, the faintest colours, greens and blues, seem to emanate outwards and become constant until one moves the eye or blinks. One's mind easily drifts into imagining velvet, the blue-black of a raven's wing, the sheen of ink, tar or lacquer, the infinity of outer space, and even death. But Reinhardt was also known for his famous 'via negativa', which he describes in 'Abstract Art Refuses': 'In painting for me, no fooling the eye, no window in the wall, no associations, no distortions, no paint caricaturings, no cream pictures or drippings ... no divine inspiration' (Rose 1991, 50). Yet it is difficult to keep Reinhardt's rational exhortations in mind when attempting to dampen associations and projections to do with the texture of matter, depth perception, darkness. As Reinhardt remarked about Zen painting and its vast emptiness, which sheds light on his own work and its ability to produce mind wandering, 'the viewer ends up by losing (finding?) himself in his own mind' (215).[35] So there is also a multistability that these paintings present between an affirmation of the freedom of the imagination, where there is a play of associations, and an emptying out of associations, closing them down to enter into a state of 'no-mind', *mushin*, extending into dark matter. These twists and turns of engagement with Reinhardt's monochromes arise haphazardly, without a supervisor, in nonlinear duration, and they combine in unique ways for each viewer. It is interesting that Reinhardt's paintings can be experienced as inertia, as a place devoid of rhythm and shiny reflections, because of the homogenous spectral reflectance provided by the evenness of the matt or satin surfaces. This is a fine balance between surface as a thing or object with a certain reflectance, and intangible darkness understood as depth, which could be a space of imagination or introspection.

Sylvester writes that some kinds of abstract art 'reflect a reductive, puritanical approach to painting, which seems concerned to find out how much of the traditional apparatus of art can be eliminated from painting without eliminating art: it's like a test of art's powers of survival

---

[35] He goes on to describe such paintings as '"weightless nothingness" with no explanations, no meanings, nothing to point out or pin down, nothing to know or feel. The least is the most; more is less' (in Rose 1991, 215).

in the face of denial of its comforts' (Sylvester 1997, 66). It is also a test of art's powers to resist the drift into the everyday. Matter painting, which is also a way of looking, presents an odd situation where it can be understood as something utterly mundane or it can undermine perceptual paradigms in order to make a creative leap. Cage writes: 'In Zen they say, if something is boring after 2 minutes try it for 4. If still boring, try it for 8, 16, 32, and so on. Eventually one discovers that it's not boring at all but very interesting' (Westgeest 1997, 80).

The repeat structures and rhythms of the everyday with which we drift into tasks and chores according to habit, familiarity or mindlessness produce a world where everything has its predictable place, so it is easy to go on automatic. But mind wandering displaces organised time; for a moment things might seem out of joint, and intensities and passions could rupture erratically through the everyday repeats and order, the semblance of normalcy, to give us glimpses of how we might possibly live or think differently *with things* we are observing to produce a 'clamor of being' (Deleuze 1995, 35). This is in consonance with the accepted view of what the artist Barnett Newman wanted to achieve with his 'empty' canvases, horizontal colourfields split down the middle with one or two vertical lines, his 'zip' paintings. These works are often understood in terms of an existential encounter with art, a kind of illumination that Zen Buddhism would describe as 'a sudden flashing into consciousness of a new truth hitherto undreamed of ... a kind of mental catastrophe' (Suzuki 1994, 65). The zip appears as a mirror of the body standing in front of it, a kind of primordial being-there. There is something of this in matter painting as well, but in a sense it goes even further in razing everything to the ground, to an imaginary beginning before humans learned to draw vertical lines in the sand or to diagrammatically represent up and down, left and right. In the teeming *Phénomènes* of Dubuffet, and in Smithson's sedimentation of the mind, we imagine the very beginnings of life, the archefossil, the metastabilities involved in minerals turning into mud and organic forms.

For Arnheim, a painting is processed using (a) identification and location of forms, and (b) their relations to each other, mapped onto (c) an egocentric, phenomenological vantage point where the viewer makes sense of patterns from the point of view of the body's responses to normal events in world. The logical coordinates of (a), (b) and (c) are assumed to be innate, continually triangulated to arrive at a moment of resolution, order and high art. But for Dada and Surrealist artists, for Smithson, Morris and Dubuffet, and for so many other artists involved in 'matter painting', these common-sense inferences from visual sense data are obstacles to new experience. The techniques of automatism, disorientation, mind wandering and de-differentiation are cultivated in order to overcome standards of design and so-called innate logic. Arnheim diminishes the explosive and destabilising aspects of Pollock's paintings by emphasising the artist's control and sense of design as the most important aspects of the work. Arnheim sees order as the most important cosmic principle that feeds through art, nature and the universe: 'Man's striving for order, of which art is but one manifestation, derives from a similar universal tendency throughout the organic world' (Arnheim 1971, 40) and 'orderliness is a state universally aspired to' (32). This point of view, which continues to fetter many earnest psychological studies of art, is vehemently opposed by Morse Peckham:

If art is the satisfaction of this mad human rage for order, think of Hitler's rage for order, of his 'final solution for the Jewish problem,' and shudder ... the desire for death is merely the desire for the most perfect order we can imagine, for total insulation from all perceptual disparities (Peckham 1965, 34).

From this point of view, when looking at Figure 3.15, Frank Stella's painting, *Arbeit macht frei* (1958), a strange cognitive dissonance, or

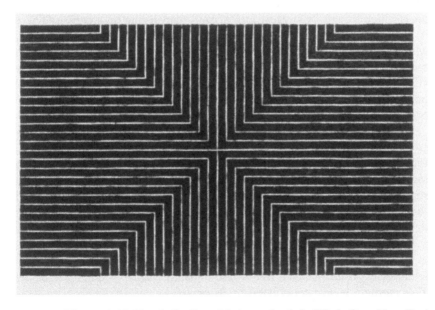

Figure 3.15 Frank Stella, *Arbeit macht frei (Work Sets Man Free)* (1967). Lithograph, 38.1 × 55.8 cm. Gemini G.E.L. Credits: Digital Image © Art Resource/Scala, Florence © Frank Stella/ARS. Copyright Agency, 2020.

metastability, arises (the lithograph is reproduced here). Enamel on canvas, the work is mainly black and overpainted with numerous repetitive pinstripe lines on the black ground. Jewish prisoners were forced at gunpoint into lines and marched under the iron arch of the concentration camp at Auschwitz, and other camps, bearing the words: '*Arbeit Macht Frei*' (work makes you free) – the false promise of freedom in exchange for labour, which led, in most cases, to their deaths. The painting is created with the artist's own 'labour', dutifully repeating lines arranged with much care and attention in mind-numbing order. The painting has been understood as a memorial eerily erased of figures, but also as a work attuned to Stella's formalist and minimalist concerns to do with 'pure' painting erased of didacticism and historical content. Cognitive dissonance is produced because the subject matter of the concentration camps is extremely difficult to 'delete' in a purely formalist interpretation. In fact, insisting on a purely formalist, rationalist interpretation of the work may come across as being callous, even unethical, and this is possibly what Stella intended.

The work produces a kind of cognitive-sensational formalism isomorphic with and expressive of specific historical conditions. The dominant compositional aspect is the division of the pictorial space into four areas which forcefully suggest four arrows converging on a centre. There is the suggestion of an 'X', arranged by the many delicate white lines, which repeat the direction of the arrows towards the centre. There is no escape for the eye beyond these dominant directional lines; the geometry virtually browbeats the viewer into submission. The clash of verticals and horizontals creates triangles that seem to pulsate at regular intervals to produce a certain discomfort.

Understood in Arnheim's terms, this work could be seen as a perfectly complete visual order. Yet it is also a 'coercive design' in the Op Art sense, forcing and anticipating a visual response of repeat rhythms leading inescapably to pointed edges in conflict with each other. Arnheim might suggest that, in the power of the centre, the solitary cross form suspended there leads to a higher resolution. But there is something in us that wants to reject this pat conclusion: the subject is the *problem* of order in all its manifestations. The painting seems to embody the idea of a rational order in art, while at the same time undermining this idea as a utopian trope by relentlessly grinding down free will associated with spontaneity and life. This moment is best articulated by Adorno:

The shock aroused by important works is not employed to trigger personal, otherwise repressed emotions. Rather, this shock is the moment in which recipients forget themselves and disappear into the work; it is the moment of being shaken. The recipients lose their footing; the possibility of truth, embodied in the aesthetic image, becomes tangible. This immediacy, in the fullest sense, of relation to artworks is a function of mediation, of penetrating and encompassing experience [*Erfahrung*]; it takes shape in the fraction of an instant, and for this the whole of consciousness is required, not isolated stimuli and responses. The experience of art as that of its truth or untruth is more than subjective experience: It is the irruption of objectivity into subjective consciousness (Adorno 1997, 244–245).

3.13   Further Developments in Dirty Rhythm

As I have mentioned, the abiding popular image of post-war American abstraction was its newness and hygienic 'immaterial' purity.[36] This in spite of Pollock's use of sand and cigarette butts in his works. Krauss notes that history sublimated Pollock into pure opticality, erasing matter, whereas in fact artists like Frank Stella were interested in the 'shrivelled materiality' of Pollock's paintings (Krauss 1993, 248). One could call the underlying psychology of these aesthetic preferences in criticism quasi-religious or 'ana-gogic', looking up to the heavens while downplaying the morbid, telluric or chthonic abyss, the scatological, the abject and filthy – the qualities associated with Bataille's *basesse* and *informe*, and with many other writers from Rimbaud and Artaud to De Sade. A significant exception to the 'clean materialism' of American abstraction was Rauschenberg's little-known *Elemental Paintings*, matter paintings executed after his trip in 1952 to Europe and North Africa with Cy Twombly. It is significant that Rauschenberg visited Rome and Burri's studio and received a painting from him, *Senza Titolo* (1952).

Rauschenberg's *Elemental Paintings* are delicate, earthy works, flecked with ochres of various hues, or covered in newspaper and fabric, over which are applied ragged or twisted patches of gold leaf. Many of these pieces were produced in the 1950s, including two white lead paintings with delicate craquelure and some red clay paintings. His *Dirt Painting (for John Cage)* (1953) included mould in a box. A unique *Growing Painting* from this period featured soil, grass and organic materials. Birdseeds sprouted in the dirt and were allowed to grow, creating a work that was literally alive. Rauschenberg explained that he regarded his artistic practice as one of collaboration *with* materials rather than 'any kind of conscious manipulation and control' (Tomkins 1968, 209). He also revealed his attitude to materials when he said that 'you begin with the possibilities of

---

[36] For a study of these artists in connection with dirt in art as a way to resist the paranoia regarding cleanliness, hygiene, foreign bodies and purity in 1950s America, see Poor (2007).

the materials, and then you let them do what they can do, so that the artist is really almost a bystander while he's working' (De Antonio and Tuchman 1984, 92). As part of this process, it was clear that his intention was to remain true to the material by letting it decay and change. The idea was to suggest that time, gravity and the constitution of the materials themselves had a role to play in mark-making. Photographer Aaron Siskind's *Chicago Series* probably gained some impetus from this work. He was a friend of Rauschenberg at Black Mountain College. Decaying materials were deeply ingrained in Siskind's visual aesthetic in cataloguing the peeling wallpaper and urban detritus of Chicago in the 1950s.

Gay artists such as Rauschenberg, Johns and others often use lessons from matter painting to reject the rhetoric of masculine abstraction. Their work is often messy and fluid, understood as a feminine trait in patriarchal texts. This queering of the purity and autonomy of macho-abstraction continues even today. For example, in New Zealand artist Shannon Novak's paintings (Figure 3.16) the artist carefully layers his acrylic

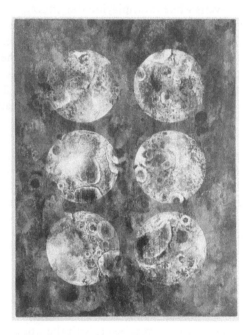

Figure 3.16 Shannon Novak, *Hadrian* (2019). Acrylic ink on board, 25 × 30 cm. © Courtesy the artist. (A black and white version of this figure will appear in some formats. For the colour version, please refer to the plate section.)

works into a pasty consistency and applies chemicals, sometimes crushed PrEP pills (pre-exposure prophylaxis, an HIV medication for people who are HIV negative), into the paint to create arbitrary textures. He then rubs or sands down the surface to reveal a deeper chromatic spectrum. The technique of soft and rhythmic frottage or rubbing yields to the more worrying chafing and grattage, a scratching that eventually breaks the skin. The artist then finishes the work with heavy varnish to achieve a mirror-like surface, where it is possible to see the faint sheen of one's own reflection. The circular form reminds the viewer of a kind of aperture one might see through a telescope or a microscope. Various associations are triggered, such as skin, pores, bacteria and even outer space, and the face of the moon. The switching between associations is rhythmic. The imagination allows the image to be perceived at a particular scale, from prehistoric alluvial valley, to diseased skin, or the microscopic ridges of cells and body fluids. The eye penetrates the depths, as it looks through the lens yet these depths are also within the viewer, who plumbs the depths of the imagination and reverie. This turns the Greenbergian aesthetic of flatness, purity and detachment into an abject, visceral and cosmic image. A variation on this work is *Skin Notations 2020 Day 25–01* (2020), the digital work on the cover of this book. The round apertures capture images of the skin, engulfed by an impenetrable darkness. The work is part of a series which documents different feelings of isolation, under lockdown in New Zealand, in response to COVID-19.

In the 1960s, a later generation of artists in America revisited the idea of decaying or disorganised matter and dirt. Beth Williamson (2017) describes how both Robert Morris and his friend and fellow artist Robert Smithson were intrigued by Ehrenzweig's concept of de-differentiation. Williamson explains that in Morris's *Threadwaste* 'the structure of the work itself (mixed, soft and indeterminate) demands a different perceptual process to be brought into play' (Williamson 2017, 157). She references Morris's writing on Ehrenzweig's 'idea of a *rhythmic alternation* between differentiation and dedifferentiation [and] chaos and resistance of chaos' (168), which result in a perceptual flux or a 'mystic oceanic feeling' – a phrase taken from Freud. Morris emphasised the process of making art, rather than a predetermined composition or plan, and in so doing suggested a different kind of observation: not the rational study of form but its gradual falling apart, not just in the perceptual moment but also in terms of deep time.

Morris's *Earthwork* (1968) was a heap of dirt on the gallery floor, riddled with industrial detritus, metal and plastic scraps and coils. His point was to question the ennoblement of the kind of matter that traditional art championed in the form of landscape. What we have in

Morris's works is a kind of dissonance: they can be read as creative and entropic, or life-affirming and morbid. Dirt and waste bring to mind Bataille's *bassesse* and *informe*, raising up the normally despised in order to countenance unspeakable, unknowable and vile matter rather than making precious its magical vitality. Since Morris, there has been a prodigious, even virulent, metastasis of trash, detritus and waste that has flowed through the corridors of art galleries the world over. The trash in the artwork and the 'trash in the mind' (Bennett 2010, 89) unite spontaneity with contingency as an anti-aesthetic, polluting the clean-cut boundary between the mind and the matter outside it. Morris's *Threadwaste* (1968) is both abstract and abject. It harks back to Duchampian parallels (sacks of coal and dust) and, like Schwitters's *merz* works, points to the ground beneath our feet. Yet even here some wonder may be found. *Threadwaste* writhes, wrapped in sparkling copper ribbons, with asphalt, steel wire and lead, all tangled up in a mass of thread waste, cotton strands of red, yellow, blue and countless other colours, the excess of human life and civilisation. Morris took something that would normally be overlooked, remnants of the textile trade, strands left to fall to the floor as a by-product of the intended outcome of refining a garment. In the middle of this chaos, Morris placed mirrors, referring to Smithson's mirror displacements. Both artists were interested in how, in a mirror reflection, matter moves the mind from perception to self-awareness. One cannot avoid the notion of matter reflecting itself. But mirrors also suggest there is something here that is worth reflecting on, something worth taking the time to contemplate. From its fine meshwork of multicoloured fibres carelessly discarded, the work emerges as an accumulation of involuntary acts where sweatshop workers have let threads fall to the floor over several days to be gathered up by the artist as a mass of countless unconscious moments, a blanket of forgetfulness.

The eyes are rhythmically jolted by the odd strip of scarlet or electric blue, or a shiny ribbon of copper. There is a scattering of rhythm even here in the trash. Morris has managed to capture the interplay of repeat forms (red thread, within many grey masses). Could this be seen as a three-dimensional Pollock? The process of observation is also subject to a general view of the mass, which disperses vision into slower and broader rhythms, waves and movements. Morris's essays suggest that observing the matter in artworks in its 'many states – from chunks, to particles, to slime' (Morris 1993, 67) deliberately undermines 'normal' object identification and achieves a scattering of form. This promotes non-conscious viewing, drifting into fluid time. Morris was well aware of the Heraclitean subtext of this kind of Process Art. In his site-specific work *Untitled (Steam Work for Bellingham)* (1971), he dug a shallow hole in the ground and arranged for steam to rise up from a bed of continually heated stones. He wrote, '[Y]ou can't put your foot into the same steam twice in this work' (Morris 2008, 62). For him, this was 'a refusal of "form" that does not, however, collapse into the sublime' (62).

3.14   Earth Art and Trends in Contemporary Art

What seems common in matter painting, forms of abstraction and various examples of Land Art, also known as Earth Art, is a relaxed attitude to art-making that gives earth, dirt, *viscosity-en-soi*, an ontological and psychological presence that is separate from the artist's control.[37] Andrew Inkpin writes that, for him, Land Art is

ontologically hybrid ... combining the artifactual, i.e. formed by human deliberation and action, it also incorporated natural aspects or elements, i.e. aspects or elements which arise by themselves and thus not determined by the artist's intervention. In such cases, the role of the artist seems best characterised not as that of a creator or maker, but as a 'gathering together' or channeling various factors – such as material, form and purpose – to yield the resultant work of art (Inkpin 2012, 263).

In his essay 'A Sedimentation of the Mind', the artist Robert Smithson attempts to describe the psychological and strongly embodied qualities of reverie that are engaged when thinking about geological matter:

The earth's surface and the figments of the mind have a way of disintegrating into discrete regions of art. Various agents, both fictional and real, somehow trade places with each other – one cannot avoid muddy thinking when it comes to earth projects, or what I will call 'abstract geology.' One's mind and the earth are in a constant state of erosion, mental rivers wear away abstract banks, brain waves undermine cliffs of thought, ideas decompose into stones of knowing, and conceptual crystallizations break apart into deposits of gritty reason. Vast moving faculties occur in this geological miasma, and they move in the most physical way. This movement seems motionless, yet it crushes the landscape of logic under glacial reveries. This slow flowage makes one conscious of the turbidity of thinking. Slump, debris slides, avalanches all take place within the cracking limits of the brain. The entire body is pulled into the cerebral sediment, where particles and fragments make themselves known as solid consciousness (Smithson 1996, 101).

Already in the 1960s, artists were struggling to describe how matter outside the mind has a 'thereness' when one is using rational extensive mental processes, but acquires an internalised 'hereness' with sensations to reach hypnagogic intensive states when thinking about the

---

[37] Two key exhibitions made explicit the interest in dirt and earth: Robert Smithson's *Earthworks*, exhibited at the Dwan Gallery in New York in 1968, and *Earth Art*, a 1969 group show at the Herbert F. Johnson Museum of Art, Cornell University.

mind-boggling aspects of deep time and uncountable granularities. It is not incidental that this kind of somatic oscillation seems to be at the root of Smithson's description, which appears to embody the drift between a human world and the primordial earth. The mental images that arise in his essay are inspired by the rhythms of sinking into the sediments of dreams and waking up out of them. We drift into the textures of dreams and images of ancient sands and primeval mud: a somatic imaginary.

In his *Site/Non-Site* works, Smithson took samples of earth from various locations in the desert, earth which had no beginning or end or containment until it was selected by the artist and piled up on the gallery floor. This displacement works on a number of levels to undermine *a priori* notions of order and place. Smithson was not interested in restating these hierarchies. He wanted us to enter a zone, a 'non-site', where the artificial boundary between mind and matter dissolves.[38] The earth used in the *Site/Non-Site* works was sometimes put into steel or glass containers, parodying archaeological finds or scientific specimens, accompanied by maps, photographs and text. All of these elements come together to resemble a museological collection, not only providing different levels of description of the earth but also displaying different phases of matter. The *Site/Non-Site* works breach the boundaries between culture and nature, earth and world, dirt and order, and various other categories. They challenge us to construct a museological order out of samples of earth, eroded by what Smithson called 'the sedimentation of the mind', the reverie induced by viewing geology as deep time. Indeed, thinking in terms of deep time while viewing Smithson's art is a special kind of reverie. The museum labels, carefully deployed, modulate the rhythmic switching between psychological states. These works arrange a situation where the earth and the world struggle:

The breakup or fragmentation of matter makes one aware of the sub-strata of the Earth before it is overly refined by industry. . . . Refinement of matter from one state to another does not mean that so-called 'impurities' of sediment are 'bad' . . . the mind and things of certain artists are not 'unities,' but things in a state of arrested disruption . . . no materials are solid, they all contain caverns and fissures.

[38] According to Timothy D. Martin, Smithson's

phenomenological practices such as his *Site/Non-Site* works did not lead to a better knowledge of the distinction between mind and matter, subject and object, but rather to an understanding of the lack of distinction between them. At this point in the process of making *Site/Non-Site* works, he encounters his 'being-in-the-world,' as Heidegger put it. And this sense of 'being' was an instance of matter's awareness of itself as matter (Martin 2011, 169).

On this view, and as with Morris's *Threadwaste*, the use of mirrors presumably engaged with the notion of matter's awareness of itself insofar as a mirror is understood as self-reflection.

Figure 3.17 Robert Smithson, *Chalk-Mirror Displacement* (1969). Sixteen mirrors and chalk. Approximately 305 cm in diameter. Through prior gift of Mr. and Mrs. Edward Morris, 1987.277. Digital Image © The Art Institute of Chicago/Art Resource, NY/Scala, Florence © Holt-Smithson Foundation/ARS. Copyright Agency, 2020.

Solids are particles built up around flux, they are objective illusions supporting grit, a collection of surfaces ready to be cracked. All chaos is put into the dark inside of the art (Smithson 1996, 100).

Smithson's samples of antediluvian sediments deposited in a gallery suggested that artists could put aside the modelling of cultural forms in the traditional manner ('sculpture'). His work points to something deeper and longer lasting: matter that existed billions of years before life itself. Not only did these particles of dirt exist before human consciousness, they might conceivably exist long after it, on loan for a short while on the gallery floor. The ancient past, the present moment and the distant future become de-differentiated as we look at the countless grains of dirt.

A literalist view of the readymade, matter painting and Smithson's *Site/Non-Site* series (Figure 3.17) might see these works simply as ordinary things, objects in everyday use and patches of earth or wall, suggesting we could just as easily go mind wandering aesthetically in a heap of gravel or a pile of junk. But what such objections fail to see is that these artworks help us to savour the psychology of reverie, 'time travel' and de-differentiation, which might otherwise fall by the wayside. For Heidegger, the artwork sets up (*aufstellen*) something that would normally be overlooked, creating a charismatic salience out of something that is ordinary in order to re-evaluate it as extraordinary. Herman de Vries, an artist who lives in a forest and arranges a field of lavender on the gallery floor, seeks to draw attention to 'the astonishing things that are taking place that we have lost the ability to see' (Gooding 2002, 55). He holds out the hope that the 'ordinary' materials released in his work provide 'an opening', not just a glimpse or a realisation, but a moment of being that produces a rift with our habitual way of subsisting. In this sense, art and matter painting in all its forms open up a clearing into a different world, where objects and piles of dirt dissolve their use value and disrupt the power we have over them. It is a clearing that is much more than a mind space or flow state, as we might venture in psychology, because in philosophical terms this is where the unprethinkable occurs. Reverie, which produces de-differentiation, is not just an ordinary response to stones and sand but can be something that is of primary importance as a way of participating in the rhythms of matter in art, different from the rhythms manufactured to make us efficient workers and consumers. This approach to engaging with art is not meant to negate the rational and cognitive understanding of it but to insist that rational understanding is not the only or most important way to engage with art.[39]

Smithson's artworks question the premise that art is simply the will to compose or order something into a finished form, representing the timeless stability of forms. He was interested in entropy as a force of change and instability. Entropy is the way in which order falls into disorder and disrepair, involving both composition and decomposition. He was mesmerised by the way in which things decompose, returning to a primordial condition, a process that propels matter forward, into the future, even as it returns to its original state, and he described this using Ehrenzweig's term 'de-differentiation'.[40] But rather than restrict its meaning to a scattering of focus, Smithson stressed how de-differentiation could be seen in terms of being and time. For Smithson, de-differentiation could be experienced as a feeling that things (sediments) are, and have been, falling apart in a fine-grained disarray for thousands of years. It can be material, temporal or psychological. This expands our understanding of 'being'. As Smithson commented in another context, describing Bangor Quarry: 'The present fell forward and backward into a tumult of "de-differentiation"' (Smithson

---

[39] There are numerous other artists who work with soil and try to raise awareness of soil degradation. For a catalogue of these artists, see Toland and Wessolek (2014).

[40] Smithson found the perfect visual example of this paradox in Mexico in 1969: the Hotel Palenque was a building that featured aspects of decay and renovation at the same time.

1996, 110). The observer participates in the scattering of particles and sediments and shares a will-less time of being. This being with matter is related to Heidegger's *Gelassenheit* or 'releasement'. It seems important that Smithson was known to own a copy of Heidegger's works and that he quotes Heidegger (Smithson 1996, 34).

Bachelard writes: 'The rocky abyss is a motionless avalanche, threatening clouds of disorder in motion' (Bachelard [1948] 2002, 143). Smithson would have enjoyed this description.[41] What Bachelard and Smithson have in common here is that imagining strata and countless corrugations sends the mind into a nonlinear state while it engages with what Bachelard called 'the temporality of granite', a 'lithochronos' (16).

In *Asphalt Rundown* (Rome, 1969), Smithson arranged for a dump truck to tip asphalt down the side of a cliff. The method of arranging this flow of materials, with some materials resisting the flow while others are overrun by it, was repeated in other sites in the same year: *Concrete Pour* (Chicago) and *Glue Pour* (Vancouver). The psychological attitude of 'letting things go' works in several ways: Smithson's releasement of traditional artistic control, allowing others (the truck driver, the workers) to release the asphalt; the matter falling down the cliff on its own, away from initial conditions; and the viewer letting go of the kind of intentional thought that would interfere with their participation in this matter-movement. For Smithson, the act of observing entropy was also a physical participation in the broader temporal process of matter falling:

The entropy of a thermally isolated system increases in the direction of the flow of psychological time ... we are dealing, not merely with an external phenomenon left to itself, but with the relationship between the external phenomenon and the consciousness which observes it (Watanabé 1969, 67).

As with the physicist Niels Bohr's idea of complementarity,

the act of observation itself increases the entropy: a shocking situation in the eyes of classical physicists since an objective physical measure is now closely dependent on the intervention of the observing subject .... In the new mechanics, entropy

---

[41] Smithson may have been familiar with Bachelard through various literary reviews. Martin (2011) writes that Smithson owned a copy of Leo Bersani 'From Bachelard to Barthes' (1967). Bersani explains Bachelard in the following terms:

Reverie is a kind of half-conscious state in which human subjects spontaneously express themselves in images of matter; either water, fire, earth or air as the 'privileged substance' for each of the four 'fundamental oneiric temperaments' which Bachelard sets out to classify in several of his works (Bersani 1967, 223).

On this view, it would be tempting to associate tendencies in Pollock, Frankenthaler and Louis with water, Masson, Still and Newman with fire, Smithson, Dubuffet and Tàpies with earth, and Rothko with air (or fire).

measures, so to speak, the degree of precision of our knowledge (Watanabé 1969, 70).

In viewing Smithson's *Asphalt Rundown*, this observer paradox is more than just the illustration of a scientific concept. It demonstrates that scientific knowledge is no longer the monopoly of science but can also inspire the expansion of art practices, happenings, photography and arrangements of matter. And it suggests that the observer is not only intrinsically involved in what is observed but is also part of the process of entropy. Entropy does not increase gradually with time

but suddenly at the moment of observation. Entropy becomes always greater after the observation than it had ever been before. Now these terms 'before' and 'after' do not refer to the flow of time, but simply to the order of the development of our cognition. We are dealing here with the direction of psychological time (Watanabé 1969, 70).

In Smithson's work, entropy and accumulation flow into each other. A stream of earth is lost at the dumping point and gained at the point where the matter falls to the ground below. But this is also the world (a truck, a cultural construct) flowing into the earth, which in Heideggerian terms is an enclosing and concealing, dampening the will of the world. The work of art reveals the struggle of these forces. As I have been arguing, dirt is born of the world; it is a cultural concept that places limits and boundaries on clean and unclean, fixing a boundary against chaos, while the earth is infinite darkness and formless abyss, an unknown fecundity: the future. In Smithson, and in other works of art I have treated here, earth turns into dirt, a construct of the world, and they seem to flow into each other.

It is interesting that in art this kind of matter can either be elevated into fecundity or denigrated as dirt. In many artworks that make room for earth, we are being asked to breach boundaries of place, site and order, to disrupt the binary of clean/polluted. On the one hand, Bataille's base materialism is tamed by geological time when Smithson's work is viewed as 'earth', and on the other hand, when viewed as 'dirt' or the pollution of a pristine lake, his work disturbs the purity of modernist formalism. This is entirely consistent with Smithson's art, which makes visible or 'frames' the process of going from order (and its cognates) to chaos (and its cognates), moving away from the ordered principles of the intellect to the disordered psychological processes of de-differentiation-as-entropy. This involves not just a metastability between dirt and earth, high and low, mind and matter, but a metastability that is a struggle for emergence, the fundamental physical process of consciousness beginning to emerge from the scattering of focus. Yet this emergence is easily scattered again among the particles. The artwork makes self-consciousness unstable and seems to point to its own 'phenomenology' of dissipating. As Jeff Malpas writes, 'the phenomenological approach is not intrinsically opposed to ontology but should rather be seen as a particular mode of ontological inquiry' (Malpas 2012, 58). This phenomenology should 'allow the phenomenon of the artwork itself to come forth, thereby allowing the artwork to exhibit its own phenomenology' (58). By extension, mind wandering does not have to be understood as simply a state of mind but rather as a state of being or becoming.

Smithson's *Spiral Jetty* (1970) is a monumental earthwork that brings together, in a large salt lake under the huge skies of Utah, earth, stones, gravel and dirt, minerals and crushed rock, coiled into a spiral. The spiral serves as a walkway into the lake, a structure reaching out into the water while curling into itself. The ambiguity of seeing earth, rocks and stone forming the image of a wave or spiral implicates the observer as a participant who can observe the whole structure from a helicopter (or in a video prepared by the artist) or step onto the work itself and lose oneself in its uncountable stones. The helicopter also spirals around the *Spiral Jetty*. On one level, the work is a rock structure that engages with the environment geologically, submerging when the water rises, re-emerging with salt crystals years later, seeming to taper off into the future while leaving in its wake large quantities of time. Seen differently, the work has mythical and oneiric qualities which suggest nonlinear time. I have mentioned, in describing Clyfford Still's work, that Bachelard held in check the logical or functionalist way of studying myths and legends because doing so trivialises myth and its power to act as a conduit for traditions of oneiric and motor imagery. The impossible and non-logical paradoxes we read about in myth, which are products of many cultural traditions that take dream images seriously, are dismissed in such studies as primitive rather than seen as a cultural and personal expression of profound fears and desires. Bachelard writes, '[C]ould legends really be passed on if they did not receive immediate support from the unconscious?' (Bachelard [1948] 2011, 152). This is the type of resonance that allows us to appreciate Smithson's *Spiral Jetty* not just as a monumental construction that illustrates the artist's connection with the great outdoors and his interest in entropy. The work can also be described in Bachelard's terms as a chthonic root or ouroboros, a buried hero who 'lives in the bowels of the Earth, living a life that while slow and drowsy, is eternal' (Bachelard [1948] 2011, 152). The connection with the oneiric is both affective and rhythmic.

Spiral Jetty is at once the great outdoors *and* the interiority of being. Encountering the path one has left behind, circling the body like a coil, one reaches the path's end and ponders cyclical repeats and eternal returns, or one sees the lit-up spaces in between, reflections of the sky and clouds, rocks and earth. These are all logical and reasonable observations, but what happens when we go deeper into the imagination, where reason is left behind? Instead of the level ground with its dull rocks, the path might suddenly seem to writhe like a serpent, turn on itself, going deeper and deeper into the sea, leaving behind 'a phosphorescent wake' (174), a petrified monument that is both perpetually turning and utterly fixed, surfacing and soon to be submerged, unfolding and folding inward. Others have felt repose at the end of the path, at the centre of the restless circumambulation. And what is repose, a feeling or a place or a kind of time? For Bachelard, 'repose is inevitably an *involutive* psyche. Turning in on oneself cannot always remain an abstraction. It can resemble the curling up on itself of a body that becomes an object for itself, that touches itself' (5). This could be a body that touches itself to affirm that it is alive and standing in the belly of the spiral: 'often we think we are only describing a world of images when we are, in fact, going down deep into our own mystery' (190).

As a path, *Spiral Jetty* is a relatively superficial human intervention in deep time, yet it hopes to add a trace in a 'lithochronos', which requires a meditative and mind-wandering frame of mind to appreciate. The work releases 'peregrination in the desert' (Lyotard 1991, 83), in Utah, walking along the spiral in a ritual manner with the earth.[42] Inevitably, this leads to comparisons with Heidegger, who describes art as the meeting place or site of conflict where the two opposing forces, earth and world, create a 'rift':

The strife that is brought into the rift and thus set back into the earth and thus fixed in place is figure, shape, Gestalt. Createdness of the work means: truth's being fixed in place in the figure. Figure is the structure in whose shape the rift composes and submits itself (Heidegger 1971, 64).

This dual dynamic is explicitly acknowledged in Heidegger's attempt to resolve the discordance between art and truth by calling attention to the double movement of revealing and concealing in the emergence of truth. This double movement is where a world opens up, where an aesthetic value or abstract concept comes forth, while at the same time the earth, the uncountable features of something particular, fades into the background. John Sallis sees this dual dynamic as 'a kind of double or

---

[42] This is a phrase that Lyotard uses to describe art-making (1991, 83). He goes on to write that art is not a game with rules that belong to cognitive mastery or that 'belong to the eye (of the prince) but to the (wandering) mind' (125). For him the aim of art 'can only be that of approaching matter' (139).

reciprocal relation between the artwork and the work-material: the art-work is made of the material, and yet also the artwork lets the material be set forth in the opening of the world' (Sallis 2008, 185).

In a letter to the poet René Char in 1971, Heidegger wrote of Cézanne: 'In the late work of the painter, the tension of emerging and not emerging has become onefold' (quoted in Petzet 1993, 143–144). It is hard not to think of this shining or shimmering in aesthetic terms. Heidegger is pointing to the peculiarity of the rift-design where the struggle between aesthetic and non-aesthetic aspects somehow becomes beautiful in itself, as a diagram of struggle.[43] The best summation of Heidegger on this point is Iain Thomson:

If, instead of trying to obtain a kind of cognitive mastery over art through aesthetics . . . we simply allow ourselves to experience what is happening within a great work of art, then Heidegger thinks we will be able to encounter the 'essential strife' in which the true work of art paradoxically 'rests' and finds its 'repose.' When we encounter the 'movement' that paradoxically rests in the masterful 'composure' of a great artwork, moreover, what we discover therein is an 'instability' that underlies the entire intelligible order, an ontological tension (between revealing and concealing, emerging and withdrawing) which can never be permanently stabilized and thus remains even in what is 'mastered' (Thomson 2011, 76).

For Heidegger, cognitive aesthetics based on idealism blinds us to the power of art by reducing it to something that merely has the appearance – the look – of a prior *eidos* or idea by which it is judged. This does not mean that Heidegger is any more tolerant of modern or romantic aesthetics that can be characterised as private and subjective. In his own way, he attempted to overcome this polarisation with the notion of 'being', a place where outward-directed perception and inner-directed introspection become continuous.

Inkpin lends support and precision to the idea that there are philosophical as well as psychological dimensions to this kind of artistic practice, writing that it is characteristic of these works to incorporate artificial and natural aspects, 'often with the intention of thematising the relationship between the two' (Inkpin 2012, 264). He notes that this tendency is found

---

[43] This could be related to the diagrammatic gestures found in Pollock, for example, and his struggle between improvisation and cultural schemata. Iain Thomson's interpretation of Heidegger's treatment of Van Gogh's painting of old shoes brings to a head a long series of disagreements between Meyer Schapiro, Jacques Derrida and others. Thomson suggests that the earthy background in this painting contains half-formed figures emerging from the earth of the paint, as do the shoes themselves, and that this emergence is both a kind of equipmental consciousness (ready-to-hand) with the more present-at-hand associated with the thematic involvement of consciousness with an aesthetic object (art, the painting). See Thomson (2011).

in Heidegger's 'treatment of artifactual (*techné*) and nature (*physis* or *phusis*) ... the first involving human action, the second independent of it' (264).[44] One could perhaps call this a kind of aesthetic quietism. However, it is the tension between learned and premeditated *techné*, and holding back from its application to allow *physis* to take its course or remain in the work as self-secluding, a gestalt or rift-design, that characterises the work itself and describes many kinds of matter painting. This complements Krauss's description of Rodin's working method, where she notes that Rodin 'forces the viewer to acknowledge the work as a result of a process .... [M]eaning does not precede experience but occurs in the process of experience itself. *It is on the surface of the work that two senses of process coincide*' (Krauss 1981b, 28–30, my italics). In other words, the material surface of the artwork can record the pattern of alternation between rational control (*techné*) and the relaxing of that control (*physis*), a tension that the viewer intuits.

Smithson's sedimentation of the mind is also a kind of painting. For the ancient Greeks, *physis* was the first creative principle, whatever it was at the dawn of the universe that changed non-existence into existence, the creative power that continues to generate such excrescences as matter-space-time, change, motion and life. For Heraclitus, *physis* denoted the totality of beings or the 'being of beings'. In Aristotle's understanding, *physis* is associated with nature and is distinct from *techné*, which is associated with art: whereas *physis* is its own source of motion, *techné* requires a source of motion outside itself. *Techné* only appears, through its own artifice, to imitate *physis*; it imitates life but is not alive in the sense of a tree or a bird. For Heidegger, *physis* is the being-ness of beings, and the earth is a manifestation of *physis*. As Michael Haar explains, this is 'not as an irruption to the fore, but as a latent thrust, a subterranean growth, a mute, concealed, nocturnal thickness' (Haar 1993, 114), which is also the material of the artwork. This suggests that an artwork is both *techné* and *physis*, insofar as the former opens up a space for the latter to be disclosed as a rift-design. But I find it interesting that *techné*, the world of technical achievement, of conscious thought, that tries to open up and utilise the earth to extract 'the truth', only ends up finding the following truth: that the earth, *physis*, is ultimately self-secluding; the earth is a limit case and defining force for the truth itself. This is consistent with the logic of Heidegger's *alēithea*: truth as a disclosure. And, interestingly, the word has within it

---

[44] The treatment of *techné* is simplified because Heidegger changes his definition of this concept in later work so that it is less oppositional and more cooperative, helping to reveal *physis*.

the root *lēithe* ('lethe'): forgetfulness or oblivion, what today we would associate with the non-conscious. The privative 'a' before 'lethe' transforms oblivion (concealedness) into disclosure (unconcealedness). Seen from a different perspective, perhaps imagistically and artistically, *alēithea* seems to suggest that oblivion is revealed as a fact in counterdistinction to the truth, almost as a moment where the two delimit each other. We realise the truth of the artwork at the moment it withdraws into its secludedness as materials, as earth.[45]

Following Inkpin, it is possible to see many Land Art works of the 1960s and 1970s as a struggle or balance between *techné* and *physis*. Many kinds of Land Art ignored traditional painting and sculpture and took art out of the gallery or museum and into the world, into deserts, lakes, industrial wasteland. Land Art often resisted the art market, creating ephemeral works or works of monumental size that could not be bought or sold. Many projects consisted of large-scale construction and engineering feats in the landscape, involving marking the earth with signs of human logic, lines or geometrical patterns, or excavating trenches in the desert, such as Michael Heizer's *Double Negative* (1969) and *Isolated Mass/Circumflex* (1968), and Walter De Maria's *Las Vegas Piece* (1969). *Double Negative* consisted of two trenches dug into the eastern edge of the Mormon Mesa in Nevada. The trenches 'underline', that is, make deeper, a natural canyon, and the dirt from the excavated trenches was dumped into the canyon. It is a wilful and emphatic human intervention in geological time, monumental and futile.

Double Negative is a massive absence and a massive presence. As with the other works discussed here, including Smithson's *Spiral Jetty*, the natural ecology, the deep time of geology, is juxtaposed with human mark-making on a megalomaniacal scale, reducing the human viewer to the perspective of an ant – or a god, if one looks at the *techné* of the aerial photographs, which is how most people view these works. These monumental works seem Janus-faced, one face looking back to ruins, ghost towns and abandoned civilisations, reminiscent of the Nazca Lines in Peru (c. AD 200–600), photographs of which had been published widely in the 1960s. Varnedoe notes how Smithson's *Spiral Jetty*, which also engages with the top-down aerial aesthetic when seen from a helicopter, resembles the spiral forms found in the Nazca Desert (Varnedoe 2006, 161). The other face seems to look to the future, creating new landmarks, the post-apocalyptic rewards of hubris. This is also a reflection of the

---

[45] There seem to be parallels here with Fried's absorption. This is premised on the logic that the painted figure's sinking into self-secluded oblivion provides the viewer with a truth limit even while sinking into their own self-seclusion (and obliviousness to being watched) while looking at the painting. See Fried (1988).

anxieties of the age, as expressed in the last scene of *Planet of the Apes* (1968) with the astronaut played by Charlton Heston's cry of anguish as he stumbles upon the remnants of the Statue of Liberty buried in sand in a canyon along the Colorado River. Undoubtedly some Land Art projects raised awareness of human destruction on a mass scale. Walter de Maria's *Mile Long Square Drawing* (1969) was situated in the Mojave Desert, not far from the main nuclear testing site in the United States from 1951 to 1992, where close to 1,000 nuclear tests were carried out.

Remote and often bizarre, these works are seldom visited and there is an air of abandonment about them. Much like abstraction itself, they empty the visual field of the social and the embodied. Today these works seem like self-fulfilling prophecies, pointing to the tendency of capitalist production to exhaust itself on follies and visions while participating in those very processes. In their own way these works bring together aspects of earth and world, using displaced objects or construction processes in the vast expanses of the wilderness to accentuate a sense of scale by framing the land, or the earth, while at the same time bringing into stark contrast human will, rationality and technology against the background of the earth, which is oblivious to reason, resting in itself. It could be argued that it is mind wandering that runs through these different works, despite the great variation in scale and duration. Both Land Art and matter painting frame non-rational content which helps to produce rhythms and associations that are also non-rational, but in Land Art it is possible, literally, to enter the work of art, to allow the mind and feet to wander in a journey that takes days rather than minutes. This brings together the notion of different kinds of wandering: mind wandering that takes place for short periods and peregrinations in the desert that we associate with the visionary or mystic, both kinds relinquishing the rational will.

Extending the human history of walking, from foraging and migration to philosophical peregrinations, are Richard Long's engagements with the earth. Since the mid-1960s, Long has made countless walks throughout the world. His works are often ephemeral, captured by photography, maps and drawings. He explains:

Each walk followed my own unique, formal route, for an original reason, which was different from other categories of walking, like travelling. . . . Thus walking – as art – provided a simple way for me to explore relationships between time, distance, geography and measurement. These walks are recorded in my work in the most appropriate way for each different idea: a photograph, a map, or a text work. All these forms feed the imagination (Long n.d.).

Rather than creating mammoth sculptural projects in the desert that express human will and technical mastery, or their futility, Long suggests that the land, the earth, can be itself while humans tread gently, making only sketchy marks that are destined to leave no trace, just like the human species itself perhaps. Long's works can be seen as minimalist, but they are also ancient in their appeal to the human body and rituals of marking place and mapping, as with the songlines of Indigenous Australians. Viewing ancient routes on a map, we imagine the time it takes to walk and traverse vast distances, and when we walk these routes we imagine those who have walked there before us, making immanent a sense of ritual. The design of these walks and the series of maps and photographs that document them are part of a deliberate, mindful mapping structure in the domain of *techné*, as is the technical mastery of the photography, which records only absences, as we see in Michael Heizer's *Double Negative*. At the same time, the peregrinations in the desert, the chaotic itinerancy through the great outdoors, are performances that are in contact with the earth, connected to *physis*, a connection through mind wandering that has become fully embodied to achieve what Heidegger would call *Dasein*.

Judy Chicago's *Atmospheres* were sited in various places in the Californian desert in the late 1960s and consisted of the artist producing smoke works, sometimes sitting on the ground, surrounded by plumes of coloured smoke and sand. There are various photographs documenting this work. It brought together aspects of performance art, Process Art and Land Art, as well as aspects of mythical thought and ritual. In Land Art in particular, there was an implicit tendency to regard the pioneering spirit, the adventurer, terra incognita and conquest, the founding of cities, as a masculine and patriarchal ideal. Chicago's works can be seen as an intervention in the male-dominated art scene of the 1960s, challenging the privilege male artists assumed in their relationship to the land. Chicago's sense of the fugacious and ephemeral aspects of the atmosphere, using smoke to immerse the body in a visionless landscape, questions the optical mastery of modernism and the pre-eminence of painting. Chicago provides her own distinct perspective on abstraction as immersive and oneiric, while pointing to the erasure of women from the art historical record of pioneering American abstract art.

An important artistic practice in these years, which complements Chicago's intervention and also points to the erasure of women, was Cuban diaspora artist Ana Mendieta's *Silueta* series (1973–1980), consisting of an outline of her body pressed into the mud, into the *physis* of the grasses and the cracked earth. As Anne Raine writes, referring to the silhouette in these works:

Its boundaries are vague and subject to immanent dissolution: at any moment, the flowers will scatter or decompose, the mud or sand will wash away, the flames will burn out, the figure will come to life and spirit itself out of the frame – or a second glance will reveal what appeared to be a human form as a momentary trick of light and shadow, a self-projection onto a chance formation of earth or wood . . . . Both sensuous and schematic, the anthropomorphic outlines simultaneously insist on the presence of the body and mark its almost palpable absence, like the chalk drawings used by police to mark the position of an absent corpse (Raine 1996, 239).

Only a commentator who is able to delve into the resources of the imagination through mind wandering can arrive at such an impressive set of insights, both philosophical and emotional, about Mendieta's work. The artwork opens up a truth: the very material of which it is made (the earth) is self-secluding. Seen in this way, Mendieta's outlines in the earth disclose the truth that the earth exists for itself and not for us. Yet the artwork is able to create a living metastability across this divide, and it seems important that the body appears to lie down with the earth and dissolve in its sediments.

Other artists have explored the body in direct contact with the earth. Dennis Oppenheim used his body to build temporary bridges or lanes across the dunes, the work becoming an endurance struggle with earth, as in *Parallel Stress* (1970). Charles Simonds's *Mythologies, Birth* and *Body-Earth*, made in the 1970s, were films that feature the artist naked, crawling in the dirt. In Terry Fox's *Levitation* (1970), the artist sat in dirt for six hours, creating a mould, leaving this imprint as the artwork. In Claes Oldenburg's *The Hole* (1967), the artist hired gravediggers to work for fourteen hours digging a hole in the ground in Manhattan, and the hole was then refilled. The work created an uncanny and oneiric regressive situation where gravediggers buried themselves in a Sisyphean task.[46] Although very different from Dubuffet's 'rehabilitation of the dirt', these works peel back the contemporary world to look at something much older that is in our midst. In burying their bodies in the earth, getting closer to it, these artists undermined the hygienic and aesthetic distinctions between dirt and earth and reconnected with ancient myths about the earth as the source of life and the receiver of death, revealing the earth to be as powerful a figure in our oneiric wandering as in our daylight existence.

Kazuo Shiraga's *Challenging Mud* (1965) also explores these themes, ritualising the cultural memory of originary struggles between the earth and the human body. Shiraga was associated with the Gutai group of artists influenced by Zen Buddhism. The term 'Gutai' is derived from '*gu*', meaning tool or technique, and '*tai*', the body, often translated as 'embodiment' or 'concreteness'. The artist was familiar with the Zen principle of becoming

---

[46] John Sturgeon's work *Narkose* (1992) can be seen in this tradition where the artist buried himself in a shallow grave.

one with the medium and strived in his performances to reveal the non-dualism of mind and body, matter and embodied action, high and low, purity and dirt. His *Challenging Mud*, which seems like an Augean task of endurance, involved repeated actions of thrashing limbs and twisting torso against the resistance of the earth. The aim was not to demonstrate the triumph of the human will and rational control over matter, what Simondon would have called the 'hylomorphic mindset'; to do so would have complemented the Expressionist paradigm, where the material is used as a tool to articulate human emotions or the ego. Instead, what is revealed is the power of the earth, its ability to absorb human challenges, to affect their fate and determine the human course of action. The churning and clammy mud, the danger of being submerged by the pull of gravity, of being sucked into the blind earth, the expressions of pain and struggle, seem both nightmarish and dangerously real. The energy expended on trying to keep the body upright in deep mud, grit, cement and twigs emerges as a life force or a birth. In step with the larger Gutai movement, which was part of his artistic and cultural milieu, Shiraga maintained that the subconscious and the sensuous were closer to matter than rational thought and could become a physical admixture with matter through the non-logical act of immersing the body in the earth.[47] It was the body that joined the earth to the subconscious by exhausting conscious thought through repeated action in the mud, a kind of full-bodied automatism. The work also complements the Heideggerian understanding of the strife between the enclosing earth and the disclosing world revealed in the work of art. In Shiraga's *Challenging Mud*, it is the force of the artist's own stamina within the energy-sapping mud that leaves a signature of the encounter between these different forces, concealing and revealing, pushing and pulling, active and passive, composing and decomposing. The performance arises from this multistability, a physically enacted struggle that leaves a material trace in the mud.

Similar developments with underlying synergies were taking place in Europe in the 1960s. The visceral chiasm of dirty matter, blood, urine and bodies was emphasised through action and motion. For Vienna Actionism, the freeing of materials was a way to 'set thoughts free':

Just as 'material thinking shall free human products from their thing-character,' one might suggest, Feminist [Vienna] Actionism shall free men's products, that is, women, from their thing-character. Just as action aims at achieving the unity of actor and material, perception and action, subject and object, Feminist Actionism seeks to transform the object of male natural history, the material 'woman',

---

[47] Williamson suggests that for Ehrenzweig mind and matter join via the 'loss of boundaries between self and not-self [that] precipitates a death of sorts where ego integrity is breached' (Williamson 2017, 171).

subjugated and enslaved by the male creator, into an independent actor and creator, subject to her own history . . . . Feminist Actionism can be traced back to Tachisme and Surrealism, which, in its techniques of automatism, articulates the repressed unconscious (Export 1989, 71).

Comparable with Vienna Actionism's Dionysian expression and base materialism is Carolee Schneemann's *Meat Joy* (1964), where she staged a mock orgy of performers writhing on the gallery floor with butchered chicken and fish. The video is mesmerising and disturbing, a fleshy mass of arms and legs rhythmically intertwined, the performers simulating copulation, writhing and caressing each other along with the fish and chicken. Schneeman's work was the exact opposite of Kantian detachment and high-minded intellectualism.

Roger Callois reminds us of the possibility – or necessity – of destabilising perceptions which need to be destroyed, to 'inflict a kind of voluptuous panic upon an otherwise lucid mind' (Callois 1958, 130). Yet in these works what seems most alarming is the undeniable fact that it is not human will that has time on its side, that human will is but a brief expenditure of energy, a flare in the infinite night sky.

In the work of Lynda Benglis, we see the biomorphism of Surrealism and elastic processes but with a surprising phenomenology of formlessness and anti-form emphatically thrown on the floor in an act of iconoclasm. Figure 3.18, *Contraband* (1969), is an early work where Benglis mixes bucketsful of liquid latex rubber with bright Day-Glo pigments. The dramatic gesture of releasing the matter onto the floor with expansive movements of the arm while pivoting the body stretches the material to fill a large space, extending matter and time with glutinous rhythms. But this matter, this energy, settles and solidifies into a kind of sculptural memorial of some sort of accident, to be scraped off the floor and made into other sculptures. One of the abiding images of Benglis is a photograph which shows her flinging a pot of heavy, viscous latex, half liquid, half paste, onto the gallery walls and floor to create a deliciously marbled slick. This was a clear reference to the art historical narrative of 'heroic' action painting, which overshadowed women's embodied actions, particularly those related to painting and sculpture. The subtext in popular accounts of Abstract Expressionism was that physical strength and mental struggle were directed outwards through the bold, uninhibited, sweeping gestures of well-muscled limbs upon the grand stage of the Abstract Expressionist canvas. According to traditional patriarchal characterisations of feminine traits, meanwhile, women are supposed to use limbs and movements with delicate and circumscribed gestures, directed inwards.

Figure 3.18 Lynda Benglis, *Contraband* (1969). Pigmented latex. Overall (irregular), 7.6 × 295.3 × 1011.6 cm. Purchased with funds from the Painting and Sculpture Committee and partial gift of John Cheim and Howard Read. Inv. N.: 2008.14. Digital Image © Whitney Museum of American Art/by Scala © Lynda Benglis/ARS. Copyright Agency, 2020. (A black and white version of this figure will appear in some formats. For the colour version, please refer to the plate section.)

This assumption can be seen in Clement Greenberg, who dismisses the 'feminine' aspects associated with handicraft and housework. Georgia O'Keeffe's paintings, for example, are singled out as having a 'lapidarian patience', with the artist 'trimming' and 'polishing' her paintings for the 'embellishments of private fetishes' (Greenberg 1988, 87), phrases suggesting hesitant gestures and repetitive movements associated with housework, needlework or polishing.[48] Clearly, spontaneity and uninhibited action were the domain of the he-man.

As I have mentioned with regard to Aristotle, there are long-standing traditions in the history of hylomorphism which equate matter with the

---

[48] Jones points out that this may not have just been gratuitous sexism on Greenberg's part (Jones 2005, 58–59). He was impressed by Irving Babbitt's writing, which extolled the virtues of masculine drawing, art and male virility over and above the cultural and moral weakness represented by women. This contempt for women was not far removed from that expressed by the Italian Futurists.

feminine. Bachelard discusses the psychological states involved in the dynamics of opposition to the resistant matter of the earth, a dynamic he describes as 'percussive' (Bachelard [1948] 2002, 33), but this 'will' is also sexual and penetrative. Here it is possible to see Bachelard's dualism of masculinity and the active principle asserted against the earth, as opposed to femininity and the passive principle associated with the earth becoming soft and wet.[49] Krauss offers a significant observation on photographer Cindy Sherman's *Untitled* works (1985–1989), which depict formless puddles of slime, decaying food, vomit and detritus. Bois and Krauss point out that the idea of the formless and viscous is often used as a vehicle for misogyny in traditional patriarchal structures that associate blood, fleshiness, vomit, bodily fluids, with female biology (Bois and Krauss 1997, 240). But gender essentialism is made somewhat androgynous in the imagination of mud or paste (*pâte*).

For Bachelard, such indeterminate consistencies create an interesting 'nausea in the hands' (Bachelard [1948] 2002, 87), which Sartre discusses in *Being and Nothingness*, with implications of an existential crisis involving the modalities of viscosity, and which so easily becomes the conscious mind's symbol for the ungraspable unconscious. The inability of the hands to control or master a substance becomes an experience of failure, not only a lack of dexterity but also a falling short of the mastery required to assert clear boundaries between pure and impure, totem and taboo, cleanliness and pollution. Viscosity and oozy latex muddy the waters, creating what Bachelard calls *viscosity-en-soi*.

Benglis's art snags historical and philosophical narratives and mythopoeic traditions and transforms their underlying prejudices into a new mythmaking. Her work is not only an effective intervention in a biased, patriarchal history of art which marginalises women, it is also highly experimental in terms of artistic practice. The thick liquid latex she uses, which eventually coagulates, is scraped off the floor and 'elevated' into sculptural form. The material retains the evidence of the chromatic marbling of contingency (or complexity), while the sculpture points to form making and aesthetic decisions. The process as a whole physically smears the mythical and essentialist boundaries between a fluid and passive feminine principle and a discrete active masculine detachment. In these performance works, *viscosity-en-soi* emerges, insofar as it is unintentional in its marbled details. Benglis also undermines the distinction between painting, sculpture and

---

[49] It is interesting that, in talking about her work *Contraband* (1969), Benglis also states that 'I think women are very involved with the ebb and flow of life and materials. . . . I think art is involved with surfaces and chemicals and mixtures and questions about matter and material' (Benglis 2009).

performance, flatness and three-dimensionality, water and earth. Another way of thinking about biomorphism and abstraction is as the stretching of objects into virtually unrecognisable continuities across the form/matter divide.

In *Contraband*, the colours swirling within the thick oozing latex are said to have sparked memories in the artist's mind of toxic oil floating on the water near a marsh in Louisiana which she used to visit as a child. The mind-wandering reverie here is a strange admixture of nostalgia and nausea. Bachelard notes how, for Edgar Allan Poe, the water is slow, gentle and silent as oil, 'a sort of plastic mediator between life and death ... this syntax of becoming ... this triple syntax of life, death and water' (Bachelard [1942] 1999, 12). Benglis plays her part in a renewed mythology as raw material for future dreams.

3.14.1 Physis *and the Earth*

A gentle and implicit aspect of matter painting, abstraction and Land Art is the engagement with the regenerative power of the earth, its ability to renew itself, reconnecting with ancient traditions of *physis*. In such works, we see how the detached and analytical world of *techné* allows *physis* to emerge. As Lack suggests:

[Heidegger] understands matter in terms of earth . . . . Matter as earth is that out of which phusis arises and to which it returns. Earth is a source of life as well as a site of depth and withdrawal. Seen in relation to a tree, for example, the earth is the nourishing source as well as the repository of the fallen giant which will decompose, returning to its source. If we conceive of matter in this fashion, we will not be as quick to see the 'earth' as a mere stockpile of resources (Lack 2014, 54).

From the 1970s, developments in Land Art were aligned to a growing ecology movement that arose in response to an increased awareness of widespread environmental pollution. We begin to see this sensibility with pioneering works by German artists Hans Haacke and Joseph Beuys, and the Hungarian artist Agnes Denes. In Joseph Beuys's *Bog Action* (1971), the artist ran amok through the mud. Many of his projects were staged 'actions', used to address concerns about urban ecology. *Overcome Party Dictatorship Now* (1971) was a reforestation project, while his *7000 Oaks* (1982–1987) was an urban revitalisation project in Kassel, Germany. He enlisted university students to plant oak tree saplings with the help of community donations.

In Haacke's *Grass Cube* (1967), followed by *Grass Grows* (1969) and, later, *Gerichtetes Wachstum (Directed Growth)* (1972), the artist took growing plants or tufts of grass into the gallery space, inserting *physis* into the sterility of minimalist formalism to suggest that the sculptural could also be living material. *Grass Cube* raises the *physis* over and above the *techné* of the modernist cube of 'idealist' transparency. *Rhine Water Purification Plant* (1972) demonstrates Haacke's interest in visualising and actualising material systems and processes. The artist filled polluted sewage water from the Rhine into a square tank in a museum and purified the water using charcoal so that it could become clean enough to provide a habitat for goldfish. A hose carried wastewater to the garden outside. As is typical of Haacke's works, this installation pointed to a larger economic and ecological problem, the pollution of the Rhine, in great contrast with the fish, a micro-ecology sustaining life even in the sterile environment of the museum. In such works, Land Art morphs into eco art, literally giving centre stage to *physis* – plants and animals – by seeding the arid zones of concrete minimalist structures with grasses and living things.

Another work that continues this line of questioning is Agnes Denes's *Wheatfield – A Confrontation* (1982), where the artist planted and harvested a two-acre field of wheat on a landfill site in New York City. It is a monumental work that brings the earth back into the zone of the world, a symbolic gesture of transforming a wasteland and songs of death into ears of wheat in downtown Manhattan. In such works, the *techné* is rendered almost invisible in order to facilitate the *physis* entangled with the 'mute, concealed, nocturnal thickness' (Haar 1993, 114) of earth. Swamped by the toxic by-products of the concrete gridded everyday, nothing could be more simple, elegant and life-affirming.

We see whispers of something similar in later works. Olafur Eliasson's *Moss Wall* (1994), for example, is made of

[r]eindeer moss (Cladonia rangiferina), a lichen native to the northern regions (including Iceland), [which] is woven into a wire mesh and mounted on the wall of a gallery. As the lichen dries, it shrinks and fades; when the installation is watered, the moss expands, changes color again, and emits a pungent odor (Eliasson n.d.).

This work, several metres high and long, takes landscape and abstract art to a new level. A living specimen installed in a museum or gallery setting, it seems a wretched vision of what might come to pass – that there may come a time when we will only know the vastness of forests and meadows through tableaus in theme parks. As with matter painting, here too art reminds us of what is vital, if we are able to see through the world's constructs which obscure it.

There is something liberating about the simplicity of these works. I see something of this in Dubuffet's idea that art is a kind of madness or desire to be wild:

[T]he only flowers I like are wild flowers. Orderly gardens make me nervous .... I feel a sharp curiosity for everything that does not emanate from man, in which man has not intervened ... wild places, wild animals ... and for my interest in worlds very different from that of man – in particular the mineral world. As for human beings, it is also their wildness that I am fond of (Dubuffet et al. 2006, 153).

On this view, it is important to see matter painting, abstract art and Land Art as examples where artists have, within their intentional, artful activity, spared as much of this wildness as possible. 'I am not a great believer in the laws concerning the nature of art', Dubuffet says. As soon as such a law is proclaimed, 'I immediately experience an intense desire to infringe it' (Dubuffet et al. 2006, 134). It is this spirit of preserving some of the 'wildness' in art that I have been concerned to disclose in this book.

3.14.2   The Earth and History

The artist Hans Haacke produced an important corrective for those who romanticise the earth as a sacred force that gives birth to a nation, the nativist's privileged connection between blood and soil. Haacke explains that one of the motivations behind his work *Der Bevölkerung* (The Population) (2000) was his experience of visiting the Reichstag building in Berlin in 1984 and seeing poor immigrant children playing in front of it. He felt that the inscription on the building, '*Dem Deutschen Volke*' ('To the German people'), placed there in 1916, excluded these children. He explains that the phrase 'implies a mythical exclusive and tribal unity ... it is a notion of racial purity ... and it is this concept of the people as a community related by blood' (Haacke 2015, xli). Haacke quotes Bertolt Brecht as saying: 'In our times anyone who says "population" in place of people or race ... is by that simple act withdrawing his support from a great many lies' (xli). In a similar vein, Deleuze and Guattari lament:

[I]t is not always easy to be Heideggerian .... He got the wrong people, earth, and blood, for the race summoned forth by art or philosophy is not one that claims to be pure but rather an oppressed, bastard, lower, anarchical, nomadic, and irremediably minor race (Deleuze and Guattari 1994, 108).

It has been said that Heidegger's continual emphasis on the earth and the soil fails to distance itself from the Nazi ideology of '*die Blut und Boden*' ('the blood and soil'), with the people (*Volk*) defined by blood and soil and filiation.[50]

---

[50] Haacke writes of his work:

Hitler had contaminated the symbolism of soil .... If ever there had been a danger in associating the project with Nazi ideology, the locations from which the participating Members of the Bundestag brought their bags of earth has made it moot. There is earth

In an interior courtyard of the Reichstag building in 2000, Haacke installed a wooden trough with white neon letters spelling the words 'DER BEVÖLKERUNG' ('To the population'), meant as a criticism of the words *Dem Deutschen Volke* inscribed on the outside of the building. Newly elected MPs were invited to bring soil from their electoral districts to deposit in the trough. The artist dedicated the work to all who happen to live within the country's borders, whatever their ethnicity or country of origin. It is interesting that several members brought soil from historical locations; for example, one member brought a sample of soil from the Jewish cemetery in their district. Over time, the garden becomes overgrown with plants, weeds and flowers, disorderly in the midst of a highly organised place of power. Within this territory grasses grow and birds and bees do their work, opening up a space to breathe, to slow things down, where the mind is free to wander, and where even the semiotics of the work begin to fall apart.

Anselm Kiefer's large-scale painting *Walhala* (2015) references the Walhalla memorial, the German Hall of Fame, inaugurated in 1842 in Donaustauf. The memorial, named after the Valhalla of Norse mythology, is modelled on the Greek temple, an interesting contrast with Heidegger's treatment of the Greek temple, where the materials of the earth merge with the emerging identity of a historical people. But in Kiefer's work, this emergence of history and being in art is a scene of devastation. Kiefer's *Walhala* depicts exploding towers, using oil, acrylic, emulsion, shellac and lead on canvas. The artist has flung these materials onto the surface and left explosive marks. Dramatic and violent, congealed and cracked, the work appears to catalogue many painting techniques in the history of matter painting, as if to suggest that all histories must fail.

Many installation works also seem to uproot notions of being and time, place and earth. Lara Almarcegui's eerie mountains of rubble gently tumble to the ground. Many of her works resemble industrial wastelands, warehouses used for rock and gravel sorting, or the ruins of war. For the Venice Biennale in 2013, the artist excavated rubble from Sacca San Mattia, a dumping ground in Venice formed by layers of waste produced by the glass and construction industries, and added piles of rubble from the dredging of the lagoon. Almarcegui brought this detritus into the Spanish Pavilion, which was constructed from similar materials. In doing so, the artist intervenes in a cycle of construction and entropy to make explicit through art the historical site-specific conditions of Venice's geography.

---

[50] from concentration camps, from Jewish cemeteries, from places where immigrants had been murdered (quoted in Deutsche et al. 2004, 79–80).

This is not dissimilar to Urs Fisher's *You* (2007), which consists of the floor of a gallery dug up with a jackhammer and filled with dirt and rubble, ruining the temple of the gallery. The work engages with Walter de Maria's *Earth Room* (1977), but rather than creating an uncanny encounter, Fisher's work makes us aware of the façade of the built environment, with its impeccable painted walls and moulded stucco, and of what lies beneath the buffed marble floors, which in his work are catastrophically rammed with earth, the origin and destination of all civilisation. Olafur Eliasson's *Riverbed* (2014), a stream with rocks and stones, installed inside Copenhagen's Louisiana Museum of Modern Art, suggests nothing less, albeit in the form of an ironic gesture towards landscape and immersive art. Here one sees the uncanny encounter between ideals to do with nature and its utterly fake and artificial recreation – what Timothy Morton labels 'ecomimesis', a romanticisation of climate Armageddon. While Morton's stinging critique may well be valid for some eco-works, it does not stop the mind from wandering into reveries about what it is we love and what it is we have lost or stand to lose. All of these works build on the idea of deconstruction, using rubble from outside the gallery to construct a vision of disaster within it, and are reflections on Smithson's *Site/Non-Site* works.

These works also echo Heidegger's understanding of the Greek temple, rising from the stone to be revealed as the art of a historical civilisation. Yet these contemporary works, rooted in older earth works, also suggest that an act of destruction, of razing to the ground, may be more revealing of the destiny of a historical people than building new temples. The *techné* seems inverted: not to build up and create, but to create something unbuilt or falling apart, as if the *techné* is only there to frame or reveal the ground zero of the earth, *physis* resting in itself. But this resting can be troubled and tainted.

This is what the artist Herman de Vries shows us with his 'earth museum', 6,500 samples of earth collected from around the world, starting in 1978 and ongoing. Sometimes these samples are finely sieved and smudged onto paper in blocks that resemble watercolour palettes, and sometimes they are arranged on the gallery floor in a kind of museological parody. Some samples are from Chernobyl and Buchenwald, others from ancient Aboriginal sites in Australia. The earth, this thing that is non-human and yet so intricately entwined with human destiny, becomes a scientific specimen undergoing classification, but the system is undermined by associations of place that cannot be shaken off in our encounter. The earth is associated with place or becomes dirt if found in the 'wrong' place. As Mary Douglas writes, dirt is 'the by-product of a systematic ordering and classification of matter, insofar as ordering involves rejecting inappropriate elements . . . is that which must not be included if a pattern is to be maintained . . . [it is] matter out of place' (Douglas 1966, 44, 50). On this view, the earth of Buchenwald is earth but it is also dirt, matter out of place in a gallery display (a work de Vries assembled in 1995). Nevertheless, it is the kind of dirt we feel the urge to respect. As Herman de Vries explains:

I took a sackful of earth from the centre of Buchenwald from barrack 15 where they did experiments on people. I presented it on a piece of white cloth . . . the shape of a grave. We don't see very much earth any more. The earth takes everything and anything into itself, but it's still witness to the facts. There was a little button off a shirt I found in the earth – a human moment in the earth (quoted in Gooding 2002, 55).

This moment brings the viewer crashing back to earth, to the ground. It halts any drift into the oceanic feeling of losing oneself in the uncountable soil, but one does return to the soil.

Framing the earth but sweeping it up into a cosmological reverie is Figure 3.19, L. N. Tallur's *Unicode* (2011), which appears to be a rough-hewn or unfinished globe of rock or earth but is in fact made of concrete, surrounded by a ring of gold. It seems to be an encounter with a coarse-grained ball of dirt or dung, the kind a scarab beetle is supposed to roll across the sky. The golden surround, shaped into flames, is taken from traditional Shiva Nataraja sculptures.

The god is creator, preserver and destroyer of the universe. Usually, it is Shiva who occupies the space in the centre, surrounded by celestial fire – the eternal circle of time. The dance of Shiva shows us the divine body in a cycle of creation and destruction, dancing in a metastable flux between the two. In *Unicode*, this is replaced with a globe, a barren rock or clump of primordial mud, equidistant from creation and destruction. This could be the beginning or the end of life on earth, just as matter painting can be seen as the destruction of culture or the beginning of a new abstraction rising from the ashes. Whatever terms we decide on, the metastability is constant.

In the traditional sculpture, Shiva holds a hand drum in his upper right hand, the first beats of creation, while the upper left hand holds the fire that destroys the universe and the lower left hand makes the *abhayamudra* gesture that allays fear. The many arms produce a nonlinear dynamic viewing experience. The figure underfoot represents *apasmara purusha* (illusion, which deceives humanity). It is interesting that within the sculpture is folded this symbol of illusion, a concept that has deep reson-ance in many philosophical systems. In Vedantic philosophy, divine knowledge is described as a stone, and creation and created forms as

Figure 3.19 L. N. Tallur, *Unicode* (2011). Bronze and concrete, 183 × 152 × 117 cm. © Courtesy the artist. (A black and white version of this figure will appear in some formats. For the colour version, please refer to the plate section.)

images carved onto that stone, but these are in essence made of the same material. In this tradition of thought, and in the works of sculpture situated within it, there is only one true reality and one indivisible unity of consciousness and the world. In European philosophy, we are familiar with this notion through Spinoza's monism, in which the single underlying substance of all reality is what he identifies as God or Nature. Here too the idea is that on the surface we see individual, separate entities or folds, but there is an underlying realm of matter where everything is connected.

In Tallur's *Unicode*, if we look closer we see that the earth, the stone, is encrusted with coins. Unicode is the underlying 'universal' digital coding structure which allows different language scripts to be displayed correctly across different computer programs and displays. The title of the work points to an interesting tension between surface effects (how the display of scripts on our computer screens suggests pluralism) and substructure (how the underlying code is used by all the different scripts, suggesting sameness and an underlying monism). In its own way, *Unicode* struggles with matter as it rests within itself and matter as it is exploited. The work allows the viewer to switch from formless matter and mind wandering to meaningful form, earth to world, dirt to currency, and back again.

The politicisation of earth as a contested substance continues in other artists' practices. The Colombian artist Miguel Ángel Rojas exhibited his work *The Future Is Already Here – It's Just Not Evenly Distributed* at the Sydney Biennale in 2016. Made of coloured sand laid in an intricate design on the floor, it is patterned after Victorian parquet flooring, a labour-intensive work of intellectually and meticulously applying the geometrical shapes and subdivisions. The work was installed on Cockatoo Island, which has a rich Victorian history of shipbuilding and also served as a prison. Seen from behind bars, with breeze flowing through the stone room, the work seems vulnerable to the elements, in danger of being blown away. In the centre of the room, a naturalistic rock formation rises from the patterned floor, representing Aboriginal and Torres Strait Islander peoples' connection to the land, a connection that is increasingly being acknowledged with plaques and at public functions in the institutions that occupy the land of indigenous peoples. There is a bi-directional flow, or multistability, whereby the raw material of the rock seems to have been pulverised into dust and sand, only to be further refined by the artistic tradition of geometry, and yet the sand can be formed into molar entities that can be identified differently. Identification and the uncountable rotate and switch. While the rock suggests topological and irregular higher dimensions of space-time and deep time, the regular patterns of the sand are very much situated in the Euclidean geometry of the Victorian period (although the pattern itself is similar to those found in Moorish Cordoba and old mansion floors in Colombia). The work is historically specific yet has multiple origins. It seems to play with the semiotics of geometry as a mapping of the earth in the context of a specific time, the era of colonisation, while the geology of deep time points to a longer history. There is also engagement with the idea of the 'sands of time' and losing oneself in the rhythms and repeats of the work. The geometry and skill of ordering the tessellations of sand engage with ideas about casting an intellectual and human net, extending over and measuring the irregular and free spaces of the earth and its geological time. The sense of gravity, molecularisation and the contractions of the world forcing its will upon the earth are all present here. This presence is not merely conceptual, as it is plain to see that the work is uncountable and ultimately chaotic, merely a collection of fine-grained particles of sand (each of which, when observed under a microscope, would have its own lines and marks) that suggest multiple rhythms and durations of geological time, worked up into coarse-grained areas shaped by the human hand where ordered lines and spaces, differentiations and colours, suggest a set regular intelligible rhythm.

Ultimately, the artist can only provide the semblance of order, with deep time and material processes momentarily organised on a coarser level of description. It is possible to understand Rojas's work as a dramatisation of the Cartesian mind/body dualism, where the viewer entertains the fiction that viewing through windows onto geometrical complications can be a disembodied experience, whereas perhaps this was merely necessary to minimise the phenomenological involvement of bodies walking on the sand and disturbing the carefully arranged patterns on the floor. One focuses and refocuses on the countable and territorial geometry, and the uncountable and infinite grains of sand and deep time. This blinking and adjusting of focus alternates with ruminating knots and expanding consciousness.

Many of these works contest the legacy of colonialism, a world system that cut up the earth and turned it into a map, severing the ancient links of indigenous peoples with the earth by displacing them. Māori artist Star Gossage's grandmother Rahui Te Kiri was evicted by the British from Little Barrier Island in 1896, when it was declared Crown land. Artist Lisa Reihana writes that 'this local history of enforced displacement helps explain the terrain as well as the melancholic tone of Gossage's work' (Reihana 2006). Gossage (Ngāti Wai/Ngāti Ruanui) lives and works on her ancestral land in Pakiri. The painting *Pakiri Pa* (2000) is made with pigments and particles of soil taken from that land in order to paint images of the land. In this way, such an act of painting with the earth is restorative. Māori are *tāngata whenua* – *of* the land. *Whenua* means both land and afterbirth, after the custom of burying the placenta in the land.

The artist states that she does not aim to depict the landscape but rather the mood that it creates and the sense of (re)connection with the land. Her aim is to paint the *wairua* of the land. The Māori concept of *wairua* may be translated roughly as soul or spirit, which may leave the body for brief periods during dreams. It is tempting to understand Gossage's moody semi-abstract paintings as opportunities to dream with the land. The ancient primeval mud, which is literally her ancestral land, creates dreamscapes for those who spend the time to allow this releasement. This is comparable with the de-differentiation of past, present and future that Smithson describes as the 'sedimentation of the mind', plunging the entire body into an ancient lithochronos which we feel we can see in the glistening ochres. A similar cluster of ideas can be found in Lisa Reihana's insightful description of Gossage's work: 'A certain combination of tar and oil may appear as watchful eyes or wrest itself into a pair of tui or a stand of trees. This conjuring act means that the images we see are a map of the subconscious moment' (Reihana 2006). In many cultures, the earth in our midst is the thing that connects us to the ancient past and sacred time. It is ancient and prehistoric; its sediments and particles existed at the time of our ancestors and ran through their fingers and toes, existed even before life itself. With the sand, mud and earth on our skin, under our fingernails, in our art, we are in direct contact with the archefossil, with the unthinkable origins of time. This quality of imagining is special when located in the sensuous dream images of art, for it deprives rational thought of its power to detach the subject from the object, to separate deep time from a ticking clock.[51]

This quality of imagining can be experienced in terms of ritual perform- ances as well, which are understood to re-enact the efficacy of originary acts, traced by material substances, making past time and present time seem indistinguishable. A well-known performance work of this kind, *Guernica in Sand* (2015) by Taiwanese artist Lee Mingwei, consists of coloured sand on the floor composed into an image of Picasso's *Guernica* (Figure 3.20). The coloured sand is divided into distinct areas by a team of workers. Visitors are then encouraged to walk over the sand, and by the end of the day, black, white and yellow sand are mixed together. This is a clear reference to the notion of entropy, as explained by Robert Smithson, using the example of a boy hopping between black and white boxes of sand. Through such constant motion, an irreversible chaotic complexity is created (Bois and Krauss 1996, 38). Mingwei's *Guernica in Sand* is completed by a performance where the artist and his assistants rhyth- mically spin around in circles, erasing any trace of the original image. It appears as if a swirling Jackson Pollock painting is swept into existence while the image of war is swept away. The archetypal primeval chaos is then worked up again into the image of war and destruction, and the cycle of creation and destruction continues, placing the human body and its rhythmic dancelike movements at the centre of this transform- ation, which also replays the story of an eternal return.

---

[51] This kind of physical, embodied bridge between art and earth was also something Richard Long was concerned with in his later work, where he hand-painted liquid mud onto the walls of the gallery with repetitive, trancelike movements. Some works, such as *River Avon Mud Arc* (2000, Guggenheim Bilbao), are more than 10 metres high, and he needed a hydraulic ladder to complete them. His *River Avon* mud-dipped paper works are simply immersed in the sediments of the ancient river, comparable with Dubuffet's *Phénomènes*, which consisted of pressing paper directly into the muddy ground or gravel.

(a)

(b)

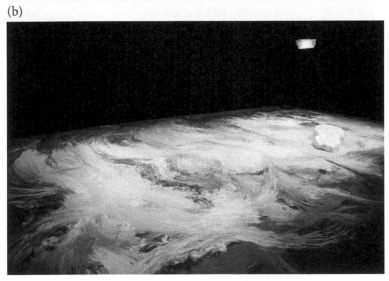

Figure 3.20 Lee Mingwei, *Guernica in Sand* (2006/2015). Mixed media interactive installation. Sand, wooden island, lighting, 1300 × 643 cm. Installation view: 'Lee Mingwei and His Relations', Taipei Fine Arts Museum, Taipei, 2015. Image: Taipei Fine Arts Museum © Courtesy the artist. (A black and white version of this figure will appear in some formats. For the colour version, please refer to the plate section.)

3.14.3   Creativity and Politico-Aesthetics

Mind wandering in psychology and art provides a deeper understanding of how subjectivity is momentarily suspended while losing a strong self-consciousness, feeling rhythmically continuous with the matter in art. Yet inevitably, we break away from this rhythm when we return to analytical thought. The emergence of the self/other distinction initiates self-reflection. There are political and existential implications to this flux of finding and losing the self.

According to Heidegger, the world thrives on 'wilful representation', the human exploitation of the earth (which includes defining it) that represents and enforces its own narrow view of what humans require and what is useful. At the root of this is *techné*, cognitive mastery, where 'nature comes to appear as a mathematical object' (Heidegger 2010, 9). Art provides us with an opportunity to get closer to materials and sensations that temporarily remove an occlusion in thought and being. This obstruction involves the encultured value we place on cognitive mastery over matter and the non-human, which trivialises mind wandering. Mind wandering is not simply a private self-indulgence, but a challenge to the assumption that only rational cognition leads to knowing. The artworks discussed here challenge our tacit understanding of mind and matter, the intellect and emotion, politics and art. The arrangement and technical execution of these works may be rational but we have a subjective commitment to what that process reveals: we care, and perhaps we also care about how the conceptual structure is expressed through the technical process and through the particularities of matter's movements and rhythms which determine the outcome of creative practice. The possibility of an isomorphism of *techné* and *physis*, of *hylē* and *morphē*, ultimately provides an important challenge to what we traditionally keep apart: political art, which we associate with conceptualism and subjectivist aesthetics associated with feeling. The dichotomy itself is loaded – is faulty.

Each artwork in its own way, with its own qualities, opens up a rift. Despite two world wars wreaking utter chaos and destruction, and more recent wars and disease across the globe inflicting death and suffering on an immense scale, creating spirals of thought in our minds about the meaninglessness and chaos of existence, the hierarchical structures of academia, art galleries and cultural institutions continue to promote the idea that art is a manifestation of a cosmic principle of order, beauty and reason that restores divinity, or at least its attributes, to 'man'. For Rudolph Arnheim, 'orderliness is a state universally aspired to' (Arnheim 1971, 32) and the most important principle in his psychology of art. This is in sharp contrast to Morse Peckham, who writes: 'After so many centuries of praising order, I think it is time to praise disorder a little ... we tend to impute order to whatever we value, even to the point of distorting perceptual data so that we see something as ordered which in fact is not' (Peckham 1965, 40–41). For Peckham, art challenges habitual rational cognition, or its supremacy, by providing a Trojan Horse opportunity to experience aspects of chaos within, to deconstruct the ordered frame that is presented to us. He writes that 'no form of human experience offered so much *disorder* as artistic experience' (39). This view shares some similarities with Ehrenzweig's interest in the dispersal of unified vision and rational categories, passed on to artists like Morris and Smithson and followed intuitively by so many abstract painters in the 1950s and 1960s. Peckham writes:

Man [*sic*] desires above all a predictable and ordered world, a world to which he is oriented, and this is the motivation behind the role of the scientist. But because man desires such a world so passionately, he is very much inclined to ignore anything that intimates that he does not have it. And to anything that disorients him, anything that requires him to experience cognitive tension he ascribes negative value (Peckham 1965, 313).

Artworks dramatise the encounter between order and disorder, the countable and the uncountable, the world and the earth. There can be an art that tends towards reason and a daydreaming art, and they can come together in the same work. When matter comes forth in artworks this becomes evident, even disorientingly so. Even today, neuroaesthetics and studies into the psychology of art presuppose that art's function is to establish order and perpetuate the 'normal' adaptive behaviour we adopt to negotiate our way through the ordinary world. This, after all, leads individuals towards an (economically) 'productive' life. Rarely is art shown to have a disorienting function that seeks to disrupt efficiency and complacency. The habitual praise for 'natural' order justifies the most conservative values that foster social repression and obscures the revolutionary anarchic possibilities of the imagination and its fundamental connection with desire.[52]

Although on the surface my interest in metastability, rhythms in the brain and mind wandering seems closely related to the project of calculating how the mind works and how we might derive value from such things as art and creativity, I have pursued these studies in order to show

---

[52] For Peckham, 'the artist's primary function is executed by offering a problem, but not a problem to be solved. He simply presents the unpredicted; he offers the experience of disorientation' (Peckham 1965, 79). But this is not just any kind of disorientation; it is disorientation of the kind that detaches us from our habitual acceptance of the authority of reason.

that science can reveal the free, natural processes of thought that flourish in spite of reason, by cooperating with chaos, not repressing it. For many of the artists and philosophers discussed in this book, the creative imagination, mind wandering and art work best when reason does not inhibit freedom. Scientific investigations into the creative and artistic mind reduce this freedom to a matter of common denominators that produce Musil's 'man without qualities'.

For many of the artists discussed in this book, 'matter' itself is understood as free – humans have it on loan only for a short period. Dubuffet states that by studying the paintings he hung in his room he was able to 'discover phenomena of every order of nature, the life of man and of all life; the mechanisms of matter and the mechanisms of mind' (Dubuffet et al. 2006, 146). This seems to express an intuition that many artists in this book explore: mind and matter are one. Artists demonstrate, not least to themselves, that the traditional mind/matter dualism fades away when one is experiencing the sustained rhythms of art-making, where neurological, muscular and kinaesthetic engagement with the texture of paint and earth joins mind and matter. It is this kind of brain-body-world and world-body-brain circular causality that disorients us, that nudges us away from the default position that human reason is at the centre of the universe. It isn't; it is only one small recursion in the far more complex dynamics of existence. Producing and viewing artworks allow us to experience this complexity.

With our busy calendars and packed schedules, how do we find the time to follow Bachelard's rhythmic and oneiric imagination of matter that sparks reveries and spontaneous moments of becoming bird, tree, crystal, dust? Artistic practices continue to materialise these questions. And it is no small victory that we are able to return to these works even after many years, to open ourselves to the depths of existence rather than drifting into the oblivion of superficial tasks, the subtle forms of rationalisation of spaces and activities that require the body to perform repetitively every day.

Existentialist philosophers have provided potent responses to world wars and the magnitude of human suffering. Their attacks on rationalist method (Hegel's 'totalitarianism of reason') were turned into acts of making art in movements such as Dada, Surrealism, *Informel* and Abstract Expressionism. Even if the motivations were only about technical innovation or aesthetic questions, the enemy was the same: excessive reason. Reason can provide solutions for a dying world. But too much of it divorces us from the things that make us desire survival. This does not mean we must force ourselves into the natural landscape and walk on all fours. It might simply mean that cognitive control associated with the executive network, considered so important by disciplinary regimes, and the fortuitous wanderings of the default mode network, squeezed into stolen moments during the 'productive' day, are equally essential for us to lead creative and fulfilling mental lives. With matter painting and abstraction, in artworks that present us with the sediments and patterns of the earth, viewers are often disoriented. Falling into infinite matter, devoid of figures, can become a rhythmic struggle where the human fails or succeeds.

References

Abbas, Niran, ed. (2005). *Mapping Michel Serres*. Ann Arbor, MI: University of Michigan Press.

Ades, Dawn. (1994). *André Masson*. New York: Rizzoli.

Adorno, Theodor. (1997). *Aesthetic Theory*. Minneapolis, MN: University of Minnesota Press.

Alloway, Laurence. (1960). *Introduction to Matter Painting*. London: Institute of Contemporary Arts.

Anfam, David. (2007). 'Transatlantic Anxieties, Especially Bill's Folly'. In *Abstract Expressionism: The International Context*, edited by Joan Marter, 51–66. New Brunswick, NJ, and London: Rutgers University Press.

Antliff, Mark. (1999). 'The Rhythms of Duration: Bergson and the Art of Matisse'. In *The New Bergson*, edited by John Mullarkey, 184–208. Manchester: Manchester University Press.

Arnheim, Rudolf. (1971). *Entropy and Art: An Essay on Disorder and Order*. Berkeley, CA: University of California Press.

Bachelard, Gaston. [1938] (1987). *The Psychoanalysis of Fire*. Boston, MA: Beacon Press.

[1942] (1999). *Water and Dreams: An Essay on the Imagination of Matter*. Dallas, TX: Dallas Institute of Humanities.

[1948] (2011). *Earth and Reveries of Repose: An Essay on Images of Interiority*. Dallas, TX: Dallas Institute of Humanities.

[1948] (2002). *Earth and Reveries of Will: An Essay on the Imagination of Matter*. Dallas, TX: Dallas Institute of Humanities.

Bagdasaryan, Julian and Michel Le Van Quyen. (2013). 'Experiencing Your Brain: Neurofeedback as a New Bridge between Neuroscience and Phenomenology'. *Frontiers in Human Neuroscience* 7: 680. www.ncbi.nlm.nih.gov/pmc/articles/PMC3807564.

Baird, Benjamin, Jonathan Smallwood, Michael D. Mrazek, Julia W. Y. Kam, Michael S. Franklin, and Jonathan W. Schooler. (2012). 'Inspired by Distraction: Mind Wandering Facilitates Creative Incubation'. *Psychological Science* 23 (10): 1117–1122. https://doi.org/10.1177/0956797612446024.

Bar, Moshe. (2009). 'The Proactive Brain: Memory for Predictions'. *Philosophical Transactions of The Royal Society B Biological Sciences* 364 (1521): 1235–1243. https://doi.org/10.1098/rstb.2008.0310.

Bar, Moshe and Maital Neta. (2006). 'Humans Prefer Curved Visual Objects.' *Psychological Science* 17 (8) (August): 645–648. https://doi.org/10.1111%2Fj .1467–9280.2006.01759.x.

Barad, Karen. (2007). *Meeting the Universe Halfway*. Durham, NC: Duke University Press.

Barrett, Estelle and Barbara Bolt, eds. (2012). *Carnal Knowledge: Towards a 'New Materialism' of the Arts*. London: I.B. Tauris.

Barrett, Lisa Feldman. (2006). 'Are Emotions Natural Kinds?' *Perspectives on Psychological Science* 1 (1) (March): 28–58. https://doi.org/10.1111/j.1745–69 16.2006.00003.x.

Barrett, Lisa Feldman, Maria Gendron, and Yang-Ming Huang. (2009). 'Do Discrete Emotions Exist?' *Philosophical Psychology* 22 (4) (August): 427–437. https://doi.org/10.1080/09515080903153634.

Barthes, Roland. (2010). *Mourning Diary*. Translated by Richard Howard. New York, NY: Hill and Wang.

Bataille, Georges. (1985). *Visions of Excess: Selected Writings, 1927–1939*. Edited by Allan Stoekl, translated by Allan Stoekl, Carl R. Lovitt, and Donald M. Leslie Jr. Minneapolis, MN: University of Minnesota Press.

Beaty, Roger E. and Rex E. Jung. (2018). 'Interacting Brain Networks Underlying Creative Cognition and Artistic Performance'. In *The Oxford Handbook of Spontaneous Thought: Mind-Wandering, Creativity, and Dreaming*, edited by Kieran C. R. Fox and Kalina Christoff, 275–284. Oxford: Oxford University Press.

Becker, Judith. (2011). 'Rhythmic Entrainment and Evolution'. In *Music, Science, and the Rhythmic Brain: Cultural and Clinical Implications*, edited by Jonathan Berger and Gabe Turow, 39–72. New York, NY: Routledge.

Becker, Madelle. (1995). 'Nineteenth-Century Foundations of Creativity Research'. *Creativity Research Journal* 8 (3): 219–229. https://doi.org/10 .1207/s15326934crj0803_2.

Belgrad, Daniel. (1998). *The Culture of Spontaneity: Improvisation and the Arts in Postwar America*. Chicago, IL: University of Chicago Press.

Bell, David. (1987). 'The Art of Judgement'. *Mind* 96 (382) (April): 221–244. https://doi.org/10.1093/mind/XCVI.382.221.

Bell, Jeffrey A. (2006). *Philosophy at the Edge of Chaos: Gilles Deleuze and the Philosophy of Difference*. Toronto: University of Toronto Press.

Benglis, Lynda. (2009). 'Whitney Focus Presents Lynda Benglis'. Whitney Museum of American Art. Published on 4 June 2009. YouTube. www .youtube.com/watch?time_continue=1&v=Yq7VkLUhY18.

Bennett, Jane. (2010). *Vibrant Matter*. Durham, NC: Duke University Press.

Bergson, Henri. [1911] (1969). *Creative Evolution*. Translated by Arthur Mitchell. Berkeley, CA: University of California Press.

    [1896] (2004). *Matter and Memory*. Translated by Nancy Margaret Paul and W. Scott Palmer. Mineola, NY: Dover Publications, Inc.

Berressem, Hanjo. (2005). '"Incerto Tempore Incertisque Locis": The Logic of the Clinamen and the Birth of Physics'. In *Mapping Michel Serres*, edited by Niran Abbas, 51–71. Ann Arbor, MI: University of Michigan Press.

Bersani, Leo. (1967). 'From Bachelard to Barthes'. *Partisan Review* 34 (2) (Spring): 215–232. http://archives.bu.edu/collections/partisan-review/searc h/detail?id=326076.

Bois, Yve-Alain, and Rosalind Krauss. (1996). 'A User's Guide to Entropy'. *October* 78 (Autumn): 38–88. https://doi.org/10.2307/778906.

(1997). *Formless: A User's Guide*. New York, NY: Zone Books, and Cambridge, MA: MIT Press.

Bonta, Mark and John Protevi. (2004). *Deleuze and Geophilosophy: A Guide and Glossary*. Edinburgh: Edinburgh University Press.

Bowden, Edward M. and Mark Jung-Beeman. (2002). 'Aha! Insight Experience Correlates with Solution Activation in the Right Hemisphere'. *Psychonomic Bulletin and Review* 10 (3): 730–737. https://doi.org/10.3758/BF03196539.

Braun, Emily, Megan Fontanella, and Carol Stringari, eds. (2015). *Alberto Burri: The Trauma of Painting*. New York, NY: Guggenheim Museum Publications.

Breton, André. (1925). 'Le Surréalisme et la Peinture'. *La révolution surréaliste* (4) (July): 26–30. https://inventin.lautre.net/livres/La-revolution-surrealiste-4.pdf.

(2002). *Surrealism and Painting*. Translated by Simon Watson. Boston, MA: MFA Publications.

Bruineberg, Jelle and Erik Rietveld. (2014). 'Self-Organization, Free Energy Minimization, and Optimal Grip on a Field of Affordances'. *Frontiers of Human Neuroscience* 8: 599. https://doi.org/10.3389/fnhum.2014.00599.

Buchanan, Ian. (2016). 'What Must We Do About Rubbish?' *Drain* 13 (1), 'Junk Ocean'. http://drainmag.com/what-must-we-do-about-rubbish.

Buchloh, Benjamin. (2002). 'Fautrier's "Natures Mortes."' In *Jean Fautrier, 1898–1964*, edited by Curtis L. Carter, Karen Butler, and Yve-Alain Bois, 63–70. New Haven, CT: Yale University Press.

Buckner, Randy L., Jessica R. Andrews-Hanna, and Daniel L. Schacter. (2008). 'The Brain's Default Network: Anatomy, Function, and Relevance to Disease'. *Annals of the New York Academy of Sciences* 1124 (1): 1–38. http s://doi.org/10.1196/annals.1440.011.

Burkeman, Oliver. (2016). 'Shuffle Your Thoughts and Sleep'. *The Guardian*. 15 July 2016. www.theguardian.com/lifeandstyle/2016/jul/15/shuffle-thoughts-sleep-oliver-burkeman.

Butler, Karen. (2002). 'Fautrier's First Critics: Andre Malraux, Jean Paulhan, and Francis Ponge'. In *Jean Fautrier, 1898–1964*, edited by Curtis L. Carter, Karen Butler, and Yve-Alain Bois, 35–56. New Haven, CT: Yale University Press.

Buzsáki, György. (2006). *Rhythms of the Brain*. Oxford: Oxford University Press.

(2019). *The Brain from Inside Out*. Oxford: Oxford University Press.

Calderone, Daniel J., Peter Lakatos, Pamela D. Butler, and F. Xavier Castellanos. (2014). 'Entrainment of Neural Oscillations as a Modifiable Substrate of Attention'. *Trends in Cognitive Sciences* 18 (6): 300–309. https://doi.org/10 .1016/j.tics.2014.02.005.

Callois, Richard. (1958). *Man, Play and Games*. Paris: Gallimard.

Carter, Curtis L. (2002). 'Fautrier's Fortunes: A Paradox of Success and Failure'. In *Jean Fautrier, 1898–1964*, edited by Curtis L. Carter, Karen Butler, and Yve-Alain Bois, 17–34. New Haven, CT, and London: Yale University Press.

Cela-Conde, Camilo J., Luigi Agnati, Joseph P. Huston, Francisco Mora, and Marcos Nadal. (2011). 'The Neural Foundations of Aesthetic Appreciation'. *Progress in Neurobiology* 94 (1): 39–48. https://doi.org/10.1016/j.pneurobio.2011.03.003.

Chave, Anna C. (1992). 'Agnes Martin: "Humility, the Beautiful Daughter .... All of Her Ways are Empty"'. In *Agnes Martin*, by Barbara Haskell, Anna C. Chave, and Rosalind Krauss, 131–154. New York, NY: Whitney Museum of American Art.

Cheetham, Mark. (1991). *Rhetoric of Purity*. Cambridge: Cambridge University Press.

(2006). *Abstract Art Against Autonomy*. Cambridge: Cambridge University Press.

Chiesa, Alberto. (2009). 'Zen Meditation: An Integration of Current Evidence'. *Journal of Alternative and Complementary Medicine* 15 (5): 585–592. https://doi.org/10.1089/acm.2008.0416.

Christoff, Kalina, Zachary C. Irving, Kieran C. R. Fox, Nathan Spreng, and Jeffery Andrews-Hannah. (2016). 'Mind-Wandering as Spontaneous Thought: A Dynamic Framework'. *Nature Reviews Neuroscience* 17: 718–731. https://doi.org/10.1038/nrn.2016.113.

Churchland, Paul. M. (2013). *Matter and Consciousness*. Cambridge, MA: MIT Press.

Clancy, Kelly. (2014). 'Your Brain is on the Brink of Chaos: Neurological Evidence for Chaos in the Nervous System is Growing'. *Turbulence* 15, 10 July 2014. http://nautil.us/issue/15/turbulence/your-brain-is-on-the-brink-of-chaos.

Clark, Andy. (1998). *Being There: Putting Brain, Body, and World Together Again*. Cambridge, MA: MIT Press.

Clark, Timothy J. (1999). *History of Modernism*. New Haven, CT: Yale University Press.

Clarke, David. (2010). *Water and Art*. London: Reaktion.

Conley, Katherine. (2013). *Surrealist Ghostliness*. Lincoln, NE: University of Nebraska Press.

Coole, Diana and Samantha Frost, eds. (2010). *New Materialisms: Ontology, Agency, and Politics*. Durham, NC: Duke University Press.

Corballis, Michael C. (2014). *The Wandering Mind: What the Brain Does When You're Not Looking*. Auckland: Auckland University Press.

Cotter, Katherine N., Paul J. Silvia, Marco Bertamini, Letizia Palumbo, and Oshin Vartanian. (2017). 'Curve Appeal: Exploring Individual Differences in Preference for Curved Versus Angular Objects'. *I-Perception* 8 (2): 1–17. https://doi.org/10.1177/2041669517693023.

Cox, Christoph, Jenny Jaskey, and Suhail Malik, eds. (2018). *Realism Materialism Art*. Berlin: Sternberg Press.

Crary, Jonathan. (2013). *24/7: Late Capitalism and the Ends of Sleep*. London: Verso.

Crowther, Paul. (2012). *The Phenomenology of Modern Art: Exploding Deleuze, Illuminating Style*. London and New York, NY: Continuum.

Csikszentmihalyi, Mihaly. (1990). *Flow: The Psychology of Optimal Experience*. New York, NY: Harper & Row.

Cummings, Paul. (1979). *Artists in Their Own Words*. London: St. Martin's Press.

Cupchik, Gerald C. (2007). 'A Critical Reflection on Arnheim's Gestalt Theory of Aesthetics'. *Psychology of Aesthetics, Creativity, and the Arts* 1 (1): 16–24. http://dx.doi.org/10.1037/1931-3896.1.1.16.

(2016). *The Aesthetics of Emotion: Up the Down Staircase of the Mind-Body*. Cambridge: Cambridge University Press. https://doi.org/10.1017 /CBO9781139169301.

Danto, Arthur C. (1983). *The Transfiguration of the Commonplace: A Philosophy of Art*. Cambridge, MA: Harvard University Press.

Da Vinci, Leonardo. [1270] (1956). *Treatise on Painting (Codex Urbinas Latinus 1270)*. Translated by Philip McMahon. Princeton, NJ: Princeton University Press.

Davis, Bret W. (2010). 'Translator's Foreword'. In *Country Path Conversations*, translated by Bret W. Davis, vii–xxii. Bloomington, IN: Indiana University Press.

Davis, Whitney. (2017). *Visuality and Virtuality: Images and Pictures from Prehistory to Perspective*. Princeton, NJ: Princeton University Press.

De Antonio, Emile and Mitch Tuchman. (1984). *Painters Painting: A Candid History of the Modern Art Scene, 1940–1970*. New York, NY: Abbeville Press.

De Bruin, Leon and Lena Kästner. (2011). 'Dynamic Embodied Cognition'. *Phenomenology and the Cognitive Sciences* 11 (4): 541–563. https://doi.org/10 .1007/s11097-011-9223-1.

De Kooning, Willem and Harold Rosenberg. (1974). *De Kooning*. New York, NY: Abrams.

DeLanda, Manuel. (2002). *Intensive Science and Virtual Philosophy*. London: Bloomsbury.

Deleuze, Gilles. (1991). *Bergsonism*. Translated by Hugh Tomlinson and Barbara Habberjam. New York, NY: Zone Books.

(1995). *Difference and Repetition*. New York, NY: Columbia University Press.

(2003). *Francis Bacon: The Logic of Sensation*. Minneapolis, MN: University of Minnesota Press.

(2004). *Difference and Repetition*. Translated by Paul Patton. London: Bloomsbury, A&C Black.

Deleuze, Gilles and Félix Guattari. (1983). *Anti-Oedipus: Capitalism and Schizophrenia*. Translated by Robert Hurly, Mark Seem, and Helen R. Lane. Minneapolis, MN: University of Minnesota Press.

(1988). *A Thousand Plateaus*. London: Athlone Press.

(1994). *What is Philosophy?* London: Verso.

Deutsche, Rosalyn, Hans Haacke, and Miwon Kwon. (2004). 'Der Bevölkerung: A Conversation'. *Grey Room* 16 (Summer): 60–81. https://doi.org/10.1162 /1526381041887448.

Dewey, John. (1934). *Art as Experience*. New York, NY: Balch.

Didi-Huberman, Georges. (2006). 'The Order of Material: Plasticities, Malaises, Survival'. In *Sculpture and Psychoanalysis*, edited by Brandon Taylor, 95–212. Aldershot, UK: Ashgate Publishing.

Dixon, Matthew L., Kieran C. R. Fox, and Kalina Christoff. (2014). 'A Framework for Understanding the Relationship between Externally and

Internally Directed Cognition'. *Neuropschologia* 62: 321–330. https://doi.org
/10.1016/j.neuropsychologia.2014.05.024.

Dixon, Matthew L., Alejandro De La Vega, Caitlin Mills, Jessica Andrews-Hanna, R. Nathan Spreng, Michael W. Cole, and Kalina Christoff. (2018). 'Heterogeneity within the Frontoparietal Control Network and its Relationship to the Default and Dorsal Attention Networks'. *Proceedings of the National Academy of Sciences* 115 (7): 1598–1607. https://doi.org/10.1073 /pnas.1715766115.

Dolphijn, Rick and Iris van der Tuin. (2012). *New Materialism: Interviews & Cartographies*. London: Open Humanities Press.

Donald, Merlin. (2006). 'Art and Cognitive Evolution'. In *The Artful Mind: Cognitive Science and the Riddle of Human Creativity*, edited by Mark Turner, 3–20. Oxford: Oxford University Press.

Douglas, Mary. (1966). *Purity and Danger: An Analysis of Concepts of Pollution and Taboo*. London: Routledge and Kegan Paul.

Drucker, Johanna. (2009). 'Entity to Event: From Literal, Mechanistic Materiality to Probabilistic Materiality'. *Parallax* 15 (4): 7–17. https://doi .org/10.1080/13534640903208834.

Dubuffet, Jean and Agnes Husslein-Arco. (2003). *Jean Dubuffet: Spur eines Abenteuers/Trace of an Adventure*. Salzburg: Salzburger Landessammlungen Rupertinum, and Bilbao: Museo Guggenheim Bilbao.

Dubuffet, Jean, Valérie da Costa, and Fabrice Hergott. (2006). *Jean Dubuffet: Works, Writings, Interviews*. Barcelona: Polígrafa.

Düchting, Hajo. (1997). *Paul Klee: Painting Music*. Munich and New York, NY: Prestel.

Dumas, Guillaume, Jacqueline Nadel, Robert Soussignan, Jacques Martinerie, and Line Garnero. (2010). 'Inter-Brain Synchronization during Social Interaction'. *PLoS ONE* 5 (8): e12166. https://doi.org/10.1371/journal .pone.0012166.

Eco, Umberto. (1989). *The Open Work*. Cambridge, MA: Harvard University Press.

Ehrenzweig, Anton. [1967] (1993). *The Hidden Order of Art: A Study in the Psychology of Artistic Imagination*. London: Weidenfeld and Nicolson.

Eifring, Halvor. (2018). 'Spontaneous Thought in Contemplative Traditions'. In *The Oxford Handbook of Spontaneous Thought: Mind-Wandering, Creativity, and Dreaming*, edited by Kieran C. R. Fox and Kalina Christoff. Oxford: Oxford University Press. https://doi.org/10.1093/oxfordhb/9780190464745.013.30.

Eliasson, Olafur. (n.d.). 'Moss Wall, 1994'. Artist's website. http://olafureliasson .net/archive/artwork/WEK101810/moss-wall.

Ellamil, Melissa, Charles Dobson, Mark Beeman, and Kalina Christoff. (2012). 'Evaluative and Generative Modes of Thought during the Creative Process'. *NeuroImage* 59 (2): 1783–1794. https://doi.org/10.1016/j.neuroimage.2011.08 .008.

Eryilmaz, Hamdi, Dimitri van de Ville, Sophie Schwartz, and Patrik Vuilleumier. (2011). 'Impact of Transient Emotions on Functional Connectivity during Subsequent Resting State: A Wavelet Correlation Approach'. *NeuroImage* 54 (3): 2481–2491. https://doi.org/10.1016/j.neuroimage.2010.10.021.

Export, Valie. (1989). 'Aspects of Feminist Actionism'. *New German Critique*, (47) (Spring–Summer): 69–92. https://doi.org/10.2307/488108.

Fazelpour, Sina and Evan Thompson. (2015). 'The Kantian Brain: Brain Dynamics from a Neurophenomenological Perspective'. *Current Opinion in Neurobiology* 31: 223–229. https://doi.org/10.1016/j.conb.2014.12.006.

Fer, Briony. (1997). *On Abstract Art*. New Haven, CT: Yale University Press.

Foster, Hal, Rosalind Krauss, Yve-Alain Bois, Benjamin Buchloh, and David Joselit. (2016). *Art Since 1900: Modernism, Antimodernism, Postmodernism*. London: Thames and Hudson.

Foster, Jonathan K. (2001). 'Cantor Coding and Chaotic Itinerancy: Relevance for Episodic Memory, Amnesia, and the Hippocampus?' *Behavioral and Brain Sciences* 24 (5): 815–816. https://doi.org/10.1017/S0140525X01280091.

Fox, Kieran C. R., Savannah Nijeboer, Elizaveta Solomonova, G. William Domhoff, and Kalina Christoff. (2013). 'Dreaming as Mind Wandering: Evidence from Functional Neuroimaging and First-Person Content Reports'. *Frontiers in Human Neuroscience* 7: 412. https://doi.org/10.3389/fnhum.2013.00412.

Franklin, Michael, Michael Mrazek, Craig Anderson, Jonathan Smallwood, Alan Kingstone, and Jonathan Schooler. (2013). 'The Silver Lining of a Mind in the Clouds: Interesting Musings are Associated with Positive Mood while Mind-Wandering'. *Frontiers in Psychology* 4: 583. https://doi.org/10.3389/fpsyg.2013.00583.

Frascina, Francis, ed. (2000). *Pollock and After: The Critical Debate*. Hove: Psychology Press.

Freedberg, David. (2006). 'Composition and Emotion'. In *The Artful Mind: Cognitive Science and the Riddle of Human Creativity*, edited by Mark Turner, 73–89. Oxford: Oxford University Press.

Freeman, Walter J. (2000). 'Brains Create Macroscopic Order from Microscopic Disorder by Neurodynamics in Perception'. In *Disorder Versus Order in Brain Function: Essays in Theoretical Neurobiology*, edited by Peter Arhem, Claes Blomberg, and Hans Liljenstrom, 205–219. London: World Scientific. https://doi.org/10.1142/9789812813398_0009.

(2001). *How Brains Make Up Their Minds*. New York, NY: University of Columbia Press.

Freeman, Walter J. and Christine A. Skarda. (1990). 'Chaotic Dynamics Versus Representationalism'. *Behavioral and Brain Sciences* 13 (1): 167–168. https://doi.org/10.1017/S0140525X00078158.

Fried, Michael. (1988). *Absorption and Theatricality: Painting and Beholder in the Age of Diderot*. Chicago, IL: Chicago University Press.

(2008). *Why Photography Matters as Art as Never Before*. New Haven, CT: Yale University Press.

Ginsborg, Hannah. (1997). 'Lawfulness without a Law: Kant on the Free Play of Imagination and Understanding'. *Philosophical Topics* 25 (1) (Spring): 37–81. www.jstor.org/stable/43154250.

Golding, John. (2000). *Paths to the Absolute*. Princeton, NJ: Princeton University Press.

Golland, Yulia, Schlomo Bentin, Hagar Gelbard, Yoav Benjamini, Ruth Heller, Yuval Nir, Uri Hasson, and Rafael Malach. (2007). 'Extrinsic and Intrinsic Systems in the Posterior Cortex of the Human Brain Revealed during Natural Sensory Stimulation'. *Cerebral Cortex* 17 (4): 766–777. https://doi.org/10.1093/cercor/bhk030.

Gooding, Mel. (2001). *Abstract Art*. Cambridge: Cambridge University Press.

(2002). *Song of the Earth: European Artists and the Landscape*. London: Thames and Hudson.

Govett-Brewster Art Gallery. (2018). *Len Lye: Heaven and Earth*. Exhibition catalogue. Govett-Brewster Art Gallery, Len Lye Centre. https://govettbrewster.com/media/uploads/2018_08/Heaven_and_Earth_Room_Brochure.pdf.

Greenberg, Clement. (1988). *The Collected Essays and Criticism, Volume 2: Arrogant Purpose, 1945–1949*. Chicago, IL: University of Chicago Press.

(2003). 'Towards a Newer Laocoon'. In *Art in Theory, 1900–2000: An Anthology of Changing Ideas*, edited by Charles Harrison and Paul Wood, 557–558. Malden, MA: Blackwell Publishers.

Grene, Marjorie. (1993). 'The Aesthetic Dialogue of Sartre and Merleau-Ponty'. In *The Merleau-Ponty Aesthetics Reader: Philosophy and Painting*, edited by Galen B. Johnson, 212–232. Evanston, IL: Northwestern University Press.

Grosz, Elizabeth. (2008). *Chaos, Territory, Art: Deleuze and the Framing of the Earth*. New York, NY: Columbia University Press.

Guattari, Félix. (1995). *Chaosmosis: An Ethico-Aesthetic Paradigm*. Bloomington, IN: Indiana University Press.

Guilbaut, Serge. (2007). 'Disdain for the Stain: Abstract Expressionism and Tachisme'. In *Abstract Expressionism: The International Context*, edited by Joan Marter, 29–50. New Brunswick, NJ, and London: Rutgers University Press.

Haacke, Hans. (2015). *Working Conditions: The Writings of Hans Haacke*. Cambridge, MA: MIT Press.

Haar, Michel. (1993). *The Song of the Earth: Heidegger and the Grounds of the History of Being*. Bloomington, IN: Indiana University Press.

Halley, Peter. (1991). 'Abstraction and Culture'. *Tema Celeste* 32–33 (Autumn): 55–60.

(1997). *Recent Essays, 1990–1996*. New York, NY: Edgewise.

Haraway, Donna J. (1997). *Modest_Witness@Second_Millennium. FemaleMan©_sMeet_ OncoMouse™*. New York, NY, and London: Routledge.

Harman, Graham. (2015). *Quentin Meillassoux*. Edinburgh: Edinburgh University Press.

Hasson, Uri, Yuval Nir, Ifat Levy, Galit Fuhrmann, and Rafael Malach. (2004). 'Intersubject Synchronization of Cortical Activity during Natural Vision'. *Science* 303 (5664): 1634–1640. https://doi.org/10.1126/science.1089506.

Hasson, Uri, Asif A. Ghazanfar, Bruno Galantucci, Simon Garrod, and Christian Keysers. (2012). 'Brain-to-Brain Coupling: A Mechanism for Creating and Sharing a Social World'. *Trends in Cognitive Sciences* 16 (2): 114–121. https://doi.org/10.1016/j.tics.2011.12.007.

Hasson, Uri, Janice Chen, and Christopher J. Honey. (2015). 'Hierarchical Process Memory: Memory as an Integral Component of Information

Processing'. *Trends in Cognitive Sciences* 19 (6): 304–313. https://doi.org/10.1016/j.tics.2015.04.006.

Haxthausen, Charles W. (2011). 'Carl Einstein, Daniel-Henry Kahnweiler, Cubism, and the Visual Brain'. *Nonsite.org* 2, 12 June 2011. https://nonsite.org/article/carl-einstein-daniel-henry-kahnweiler-cubism-and-the-visual-brain.

Hayles, Katherine N. (1990). *Chaos Bound: Orderly Disorder in Contemporary Literature and Science*. Ithaca, NY: Cornell University Press.

(2017). *Unthought: The Power of the Cognitive Nonconscious*. Chicago, IL: University of Chicago Press.

Heidegger, Martin. (1967). *Being and Time*. Translated by John Macquarrie and Edward Robinson. New York, NY: Harper & Row.

(1971). *Poetry, Language, Thought*. Translated by Albert Hofstadter. New York, NY: Harper & Row.

(1972). *On Time and Being*. Translated by Joan Stambaugh. New York, NY: Harper & Row.

(2002). 'The Origin of the Work of Art'. In *Off the Beaten Track*, edited and translated by Julian Young and Kenneth Haynes. Cambridge: Cambridge University Press.

(2010). *Country Path Conversations*. Translated by Bret W. Davis. Bloomington, IN: Indiana University Press.

Henderson, Linda Dalrymple. (1983). *The Fourth Dimension and Non-Euclidean Geometry in Modern Art*. Princeton, NJ: Princeton University Press.

Herbert, James D. (2015). *Brushstroke and Emergence: Courbet, Impressionism, Picasso*. Chicago, IL: University of Chicago Press.

Hiramatsu, Chihiro, Naokazu Goda, and Hidehiko Komatsu. (2011). 'Transformation from Image-Based to Perceptual Representation of Materials along the Human Ventral Visual Pathway'. *NeuroImage* 57 (2): 482–494. https://doi.org/10.1016/j.neuroimage.2011.04.056.

Houston, Joe. (2007). *Optic Nerve: Perceptual Art of the 1960s*. London and New York, NY: Merrell.

Hutchins, Edwin. (2005). 'Material Anchors for Conceptual Blends'. *Journal of Pragmatics* 37 (10): 1555–1577. https://doi.org/10.1016/j.pragma.2004.06.008.

Ingold, Tim. (2007). 'Materials Against Materiality'. *Archaeological Dialogues* 14 (1): 1–16. https://doi.org/10.1017/S1380203807002127.

Ingram, Simon. (2004). 'Machinic Practice in Painting.' *In Junctures*, 2 (June): 33–44.

Inkpin, Andrew. (2012). 'The Complexities of "Abstracting" from Nature'. In *Meanings of Abstract Art: From Nature to Theory*, edited by Isabel Wünsche and Paul Crowther, 255–269. London: Routledge.

Ishaghpour, Youssef. (2006). *Antoni Tàpies: Works, Writings, Interviews*. Barcelona: Ediciones Polígrafa, and New York, NY: Distributed Art Publishers.

Jackson, Robert. (2013). 'Morphism, Anti-Literalism and Presentness'. In *Speculations II: The Anxiousness of Objects and Artworks*, edited by Ridvan Askin, Paul J. Ennis, Andreas Hägler, and Philipp Schweighauser, 135–168. New York, NY: Punctum Books.

Johnston, Adrian. (2013). *Prolegomena to Any Future Materialism. Volume One: The Outcome of Contemporary French Philosophy*. Evanston, IL: Northwestern University Press.

Jones, Caroline A. (2005). *Eyesight Alone: Clement Greenberg's Modernism and the Bureaucratization of the Senses.* Chicago, IL: University of Chicago Press.

(2016). 'A Questionnaire on Materialisms'. *October* 155 (Winter): 61–63. https://doi.org/10.1162/OCTO_a_00243.

Kandel, Eric R. (2016). *Reductionism in Art and Brain Science: Bridging Two Cultures.* New York, NY: Columbia University Press.

Kandinsky, Wassily. (1977). *Concerning the Spiritual in Art.* New York, NY: Dover.

Kaplan, Stephen. (1995). 'The Restorative Benefits of Nature: Towards an Integrative Framework'. *Journal of Environmental Psychology* 15 (3): 169–182. https://doi.org/10.1016/0272-4944(95)90001-2.

Kellert, Stephen H. (2009). *Borrowed Knowledge: Chaos Theory and the Challenge of Learning Across Disciplines.* Chicago, IL: University of Chicago Press.

Kelso, J. A. Scott. (1995). *Dynamic Patterns: The Self-Organization of Brain and Behavior.* Cambridge, MA: MIT Press.

Kelso, J. A. Scott, Guillaume Dumas, and Emmanuelle Tognoli. (2013). 'Outline of a General Theory of Behavior and Brain Coordination'. *Neural Networks* 37: 120–131. https://doi.org/10.1016/j.neunet.2012.09.003.

Kemp, Martin. (2000). *Visualizations: The Nature Book of Art and Science.* Berkeley, CA: University of California Press.

Kim, Chai-Youn and Randolph Blake. (2007). 'Brain Activity Accompanying Perception of Implied Motion in Abstract Paintings'. *Spatial Vision* 20 (6): 545–560. https://doi.org/10.1163/156856807782758395.

Klee, Paul. (1964). *The Diaries of Paul Klee, 1898–1918.* Edited by Felix Klee. Berkeley, CA: University of California Press.

[1925] (1972). *Pedagogical Sketchbook.* New York, NY: Frederick A. Praeger.

Knappett, Carl. (2005). 'The Affordances of Things: A Post-Gibsonian Perspective on the Relationality of Mind and Matter'. In *Rethinking Materiality: Engagement of Mind with the Material World,* edited by Elizabeth DeMarrais, 43–51. Cambridge, UK: McDonald Institute for Archaeological Research.

Korn, Henri and Phillipe Faure. (2003). 'Is There Chaos in The Brain? II. Experimental Evidence and Related Models'. *Comptes Rendus Biologies* 326 (9): 787–840. https://doi.org/10.1016/j.crvi.2003.09.011.

Kowatari, Yasuyuki, Seung Hee Lee, Hiromi Yamamura, Yusuke Nagamori, Pierre Levy, Shigeru Yamane, and Miyuki Yamamoto. (2009). 'Neural Networks Involved in Artistic Creativity'. *Human Brain Mapping* 30: 1678–1690. https://doi.org/10.1002/hbm.20633.

Krauss, Rosalind E. (1979). 'Grids'. *October* 9 (Summer): 50–64. https://doi.org/10.2307/778321.

(1981a). *Passages in Modern Sculpture.* Cambridge, MA: MIT Press.

(1981b). 'The Photographic Conditions of Surrealism'. *October* 19 (Winter): 3–34. https://doi.org/10.2307/778652.

(1985). 'Corpus Delicti'. *October* 33 (Summer): 31–72. https://doi.org/10.2307/778393.

(1993). *The Optical Unconscious.* Cambridge, MA: MIT Press.

Kulezic-Wilson, Danijela. (2015). *The Musicality of Narrative Film.* Basingstoke: Palgrave Macmillan.

Lack, Anthony. (2014). *Martin Heidegger on Technology, Ecology, and the Arts*. Basingstoke: Palgrave Macmillan.

Lamm, Claus C., Daniel Batson, and Jean Decety. (2007). 'The Neural Substrate of Human Empathy: Effects of Perspective-Taking and Cognitive Appraisal'. *Journal of Cognitive Neuroscience* 19 (1): 42–58. https://doi.org/10.1162/jocn .2007.19.1.42.

Lampert, Jay. (2012). *Simultaneity and Delay: A Dialectical Theory of Staggered Time*. London: Bloomsbury.

Lange-Berndt, Petra, ed. (2015). *Materiality*. London: Whitechapel Gallery and Cambridge, MA: MIT Press.

Leclerc, Ivor. (2004). *The Nature of Physical Existence*. New York, NY: Psychology Press.

Lefebvre, Henri. (2004). *Rhythmanalysis: Space, Time and Everyday Life*. London and New York, NY: Continuum.

Lerner, Vladimir and Eliezer Witztum. (2015). 'The Artist, Depression, and the Mood Landscape'. *American Journal of Psychiatry* 172 (3): 225–226. https:// doi.org/10.1176/appi.ajp.2014.14091159.

Levine, Gregory P. A. (2017). *Long Strange Journey: On Modern Zen, Zen Art, and Other Predicaments*. Honolulu, HI: University of Hawaii Press.

Lewis, Marc D. (2005). 'Bridging Emotion Theory and Neurobiology through Dynamic Systems Modeling'. *Behavioral and Brain Sciences* 28 (2): 169–194. https://doi.org/10.1017/S0140525X0500004X.

Limb, Charles J. and Allen R. Braun. (2008). 'Neural Substrates of Spontaneous Musical Performance: An fMRI Study of Jazz Improvisation'. *PLoS ONE* 3 (2): e1679. https://doi.org/10.1371/journal.pone.0001679.

Liu, Siyuan, Ho Ming Chow, Yisheng Xu, Michael G. Erkkinen, Katherine E. Swett, Michael W. Eagle, Daniel A. Rizik-Baer, and Allen R. Braun. (2012). 'Neural Correlates of Lyrical Improvisation: An fMRI Study of Freestyle Rap'. *Scientific Reports* 2: 834. https://doi.org/10.1038/srep00834.

Livingstone, Margaret. (2002). *Vision and Art: The Biology of Seeing*. New York, NY: H. N. Abrams.

Long, Richard. (n.d.). 'Richard Long: Heaven and Earth, Room 2'. Tate Gallery. Exhibitions and events. www.tate.org.uk/whats-on/tate-britain/exhibition/ri chard-long-heaven-and-earth/richard-long-heaven-and-earth-1.

Loreau, Max. (1965). *Catalogue des Travaux de Jean Dubuffet. Fascicule XIX: Paris Circus*. Translated by Thomas Repensek. New York, NY: Cordier and Ekstrum.

Lundholm, Helge. (1921). 'The Affective Tone of Lines: Experimental Researches'. *Psychological Review* 28 (1): 43–60. http://dx.doi.org/10.1037 /h0072647.

Lutz, Antoine, Jean-Philippe Lachaux, Jacques Martinerie, and Francisco J. Varela. (2002). 'Guiding the Study of Brain Dynamics by Using First-Person Data: Synchrony Patterns Correlate with Ongoing Conscious States during a Simple Visual Task'. *Proceedings of the National Academy of Sciences* 99 (3): 1586–1591. https://doi.org/10.1073/pnas.032658199.

Lutz, Antoine and Evan Thompson. (2003). 'Neurophenomenology: Integrating Subjective Experience and Brain Dynamics in the Neuroscience of

Consciousness'. *Journal of Consciousness Studies* 10 (9–10): 31–52. https://p
hilpapers.org/rec/LUTNI.

Lyotard, Jean-François. (1991). *The Inhuman: Reflections on Time*. Stanford, CA:
Stanford University Press.

Malabou, Catherine. (2008). *What Should We Do With Our Brain?* New York,
NY: Fordham.

Malafouris, Lambros. (2013). *How Things Shape the Mind: A Theory of Material
Engagement*. Cambridge, MA: MIT Press.

Malpas, Jeff. (2012). *Heidegger and the Thinking of Place: Explorations in the
Topology of Being*. Cambridge, MA: MIT Press.

Marabissi, Dania and Romano Fantacci. (2015). *Cognitive Interference
Management in Heterogeneous Networks*. Springer: Dordrecht.

Marcuse, Herbert. (1978). *The Aesthetic Dimension: Towards a Critique of Marxist
Aesthetics*. Boston, MA: Beacon Press.

Marks, John. (2006). 'Introduction.' *Paragraph* 29 (2): 1–18. www.jstor.org/sta
ble/43151939.

Marter, Joan, ed. (2007). *Abstract Expressionism: The International Context*. New
Brunswick, NJ, and London: Rutgers University Press.

Martin, Kirsty. (2013). *Modernism and the Rhythms of Sympathy*. Oxford: Oxford
University Press.

Martin, Timothy D. (2011). 'Robert Smithson and the Anglo-American
Picturesque'. In *Anglo-American Exchange in Postwar Sculpture, 1945–1975*,
edited by Rebecca Peabody, 164–174. Los Angeles, CA: J. Paul Getty
Museum.

Martins, Bruna and Mara Mather. (2016). 'Default Mode Network and Later-
Life Emotion Regulation: Linking Functional Connectivity Patterns and
Emotional Outcomes'. In *Emotion, Aging, and Health*, edited by Anthony
D. Ong and Corinna E. Löckenhoff, 9–29. Washington, DC: American
Psychological Association. http://dx.doi.org/10.1037/14857–002.

Marx, Karl. [1844] (1977). *Critique of Hegel's 'Philosophy of Right'*. Translated by
Joseph J. O'Malley and Annette Jolin. Cambridge: Cambridge University Press.

Massaro, Davide, Federica Savazzi, Cinzia Di Dio, David Freedberg,
Vittorio Gallese, Gabriella Gilli, and Antonella Marchetti. (2012). 'When
Art Moves the Eyes: A Behavioral and Eye-Tracking Study'. *PLoS ONE* 7
(5): e37285. https://doi.org/10.1371/journal.pone.0037285.

Mausfeld, Rainer. (2010). 'The Perception of Material Qualities and the Internal
Semantics of the Perceptual System'. In *Perception Beyond Inference: The
Information Content of Visual Processes*, edited by Liliana Albertazzi,
Gert J. van Tonder, and Dhanraj Vishwanath. Cambridge, MA: MIT Press.

Mauzerall, Hope. (1998). 'What's the Matter with Matter? Problems in the
Criticism of Greenberg, Fried, and Krauss'. *Art Criticism* 13 (1): 81–96.

May, Rollo. (1958). *Existence: A New Dimension in Psychiatry and Psychology*.
New York, NY: Simon & Schuster.

McCumber, John. (2013). *On Philosophy: Notes from a Crisis*. Stanford, CA:
Stanford University Press.

McMullin, Ernan. (1963). *The Concept of Matter*. Notre Dame, IN: University of
Notre Dame Press.

Mehta, Ravi and Rui Zhu. (2009). 'Blue or Red? Exploring the Effect of Color on Cognitive Task Performances'. *Science* 323 (5918): 1226–1229. https://doi .org/10.1126/science.1169144.

Meillassoux, Quentin. (2007). 'Subtraction and Contraction: Deleuze, Immanence, and Matter and Memory'. In *Collapse Volume III: Unknown Deleuze [+Speculative Realism]*, edited by Robin Mackay, 63–107. www .urbanomic.com/book/collapse-3.

Melville, Herman. [1851] (2007). *Moby Dick*. London: Vintage Books.

Merleau-Ponty, Maurice. (1993). 'Cézanne's Doubt'. In *The Merleau-Ponty Aesthetics Reader: Philosophy and Painting*, edited by Galen B. Johnson, 59–75. Evanston, IL: Northwestern University Press.

Messensee, Caroline. (2003). 'Jean Dubuffet—Spur eines Abenteuers'. In *Jean Dubuffet: Spur eines Abenteuers/Trace of an Adventure*, by Jean Dubuffet and Agnes Husslein-Arco, 28–34. Salzburg: Salzburger Landessammlungen Rupertinum, and Bilbao: Museo Guggenheim Bilbao.

Michotte, Albert. [1954] (1991). 'Amodal Completion of Perceptual Structures'. In *Michotte's Experimental Phenomenology of Perception*, edited by Georges Thinès, Alan Costall, and George Butterworth, 140–167. New York, NY: Lawrence Erlbaum Associates.

Miles, Malcolm. (2014). *Eco-Aesthetics: Art, Literature and Architecture in a Period of Climate Change*. London: Bloomsbury, A&C Black.

Miller, Daniel. (2007). 'Stone Age or Plastic Age?' *Archaeological Dialogues* 14 (1): 23–27. https://doi.org/10.1017/S1380203807002152.

Minissale, Gregory. (2013). *The Psychology of Contemporary Art*. Cambridge: Cambridge University Press.

Minturn, Kent. (2007). 'Greenberg Misreading Dubuffet'. In *Abstract Expressionism: The International Context*, edited by Joan Marter, 29–50. New Brunswick, NJ, and London: Rutgers University Press.

Mitchell, Timothy. (1977). 'Bergson, Le Bon and Hermetic Cubism'. *Journal of Aesthetics and Art Criticism* 36 (2). (Winter): 175–183. https://doi.org/10.2307 /429757.

Mole, Christopher. (2011). *Attention is Cognitive Unison: An Essay in Philosophical Psychology*. New York, NY: Oxford University Press.

Morris, Robert. (1993). *Continuous Project Altered Daily: The Writings of Robert Morris*. Cambridge, MA: MIT Press.

(2008). *I Have Reasons: Works and Writings, 1993–2007*. Edited by Nena Tsouti-Schillinger. Durham, NC: Duke University Press.

Mrazek, Michael D., Benjamin W. Mooneyham, and Jonathan W. Schooler. (2014). 'Insights from Quiet Minds: The Converging Fields of Mindfulness and Mind-Wandering'. In *Meditation—Neuroscientific Approaches and Philosophical Implications*, edited by Stefan Schmidt and Harald Walach, 227–241. Dordrecht: Springer.

Mullarkey, John. (1999). 'Deleuze and Materialism: One or Several Matters?' In *A Deleuzian Century?*, edited by Ian Buchanan, 59–84. Durham, NC, and London: Duke University Press. https://doi.org/10.1215/9780822395973-004.

Murphie, Andrew. (2010). 'Deleuze, Guattari, and Neuroscience'. In *Deleuze, Science and the Force of the Virtual*, edited by Peter Gaffney, 330–367.

Minneapolis, MN: University of Minnesota Press. https://doi.org/10.5749/minnesota/9780816665976.003.0012.

Muzur, Amir, Edward F. Pace-Schott, and J. Allan obson. (2002). 'The Prefrontal Cortex in Sleep'. *Trends in Cognitive Sciences* 6 (11): 475–481. ht tps://doi.org/10.1016/S1364-6613(02)01992-7.

Nodine, Calvin F. and Elizabeth Krupinski. (2003). 'How Do Viewers Look at Artworks?' *Bulletin of Psychology and the Arts* 4: 65–68.

Nöe, Alva. (2004). *Action in Perception*. Cambridge, MA: MIT Press.

Northoff, Georg. (2012). 'Immanuel Kant's Mind and the Brain's Resting State'. *Trends in Cognitive Sciences* 16 (7): 356–359. https://doi.org/10.1016/j .tics.2012.06.001.

Papadopoulo, Alexandre. (1980). *Islam and Muslim Art*. London: Thames and Hudson.

Paulhan, Jean. (2002). 'Fautrier the Enraged'. In *Jean Fautrier, 1898–1964*, edited by Curtis L. Carter, Karen Butler, and Yve-Alain Bois, 178–187. New Haven, CT, and London: Yale University Press.

Paulson, Eric J. (2005). 'Viewing Eye Movements during Reading through the Lens of Chaos Theory: How Reading Is Like the Weather'. *Reading Research Quarterly* 40 (3): 338–358. https://doi.org/10.1598/RRQ.40.3.3.

Peckham, Morse. (1965). *Man's Rage for Chaos: Biology, Behavior, and the Arts*. Philadelphia, PA: Chilton Books.

Pessoa, Luiz. (2010). 'Emergent Processes in Cognitive-Emotional Interactions'. *Dialogues in Clinical Neuroscience* 12 (4): 433–448. www.ncbi.nlm.nih.gov/p mc/articles/PMC3117594.

Petrie, Brian. (1974). 'Boccioni and Bergson'. *Burlington Magazine* 116 (852): 140–147. www.jstor.org/stable/877621.

Petzet, Heinrich W. (1993). *Encounters and Dialogues with Martin Heidegger, 1929–1976*. Translated by Parvis Emad and Kenneth Maly. Chicago, IL: University of Chicago Press.

Pisters, Patricia. (2012). *The Neuro-Image: A Deleuzian Film-Philosophy of Digital Screen Culture*. Stanford, CA: Stanford University Press.

Polcari, Stephen. (1991). *Abstract Expressionism and the Modern Experience*. Cambridge, MA: Cambridge University Press.

Ponge, Francis. (2002). 'Note on the Otages'. In *Jean Fautrier, 1898–1964*, edited by Curtis L. Carter, Karen Butler, Yve-Alain Bois, 170–177. New Haven, CT, and London: Yale University Press.

Poor, Natasha. (2007). 'Reconsidering the Work of Robert Rauschenberg, Bruce Conner, Jay DeFeo, and Edward Kienholz'. PhD dissertation, New York City University.

Pritzker, Steven R. (1999). 'Zen'. In *The Encyclopedia of Creativity*, edited by Mark Runco and Steven Pritzker, 539–543. Amsterdam: Elsevier.

Protevi, John. (2001). *Political Physics: Deleuze, Derrida and the Body Politic*. London: Athlone Press.

    (2009). *Political Affect: Connecting the Social and the Somatic*. Minneapolis, MN: University of Minnesota Press.

Purcell, Allan T., Richard. J. Lamb, Erminielda Mainardi Peron, and Susanna Falchero. (1994). 'Preference or Preferences for Landscape'.

Journal of Environmental Psychology 14 (3): 195–209. https://doi.org/10.1016
/S0272-4944(94)80056-1.

Purcell, Terry, Erminielda Peron, and Rita Berto. (2001). 'Why Do Preferences
Differ between Scene Types?' *Environment and Behavior* 33 (1): 93–106. ht
tps://doi.org/10.1177/00139160121972882.

Raine, Anne. (1996). 'Embodied Geographies: Subjectivity and Materiality in the
Work of Ana Mendieta'. In *Generations and Geographies in the Visual Arts:
Feminist Readings*, edited by Griselda Pollock, 228–252. New York, NY, and
London: Routledge.

Redies, Christoph. (2008). 'A Universal Model of Eesthetic Perception Based on the
Sensory Coding of Natural Stimuli'. *Spatial Vision* 21 (1–2): 97–117. https://doi
.org/10.1163/156856807782753886.

Reihana, Lisa. (2006). 'Maps of Memories: The Art of Star Gossage'. *Art New
Zealand* 118 (Autumn): 42–45. www.art-newzealand.com/Issue118/star.htm.

Rimbaud, Arthur. [1871] (1966). 'Letter to Izambard'. May 1871. In *Complete
Works and Selected Letters*, edited and translated by Wallace Fowlie, 303–304.
Chicago, IL: University of Chicago Press.

Rose, Barbara. (1991). *Art as Art: The Selected Writings of Ad Reinhardt*. Berkeley,
CA: University of California Press.

Rosenthal, Mark. (1996). *Abstraction in the Twentieth Century: Total Risk,
Freedom, Discipline*. New York, NY: H. N. Abrams.

Rubin, William and Carolyn Lanchner. (1976). *André Masson*. New York, NY:
Museum of Modern Art.

Runco, Mark. (2006). 'Reasoning and Personal Creativity'. In *Knowledge and
Reason in Cognitive Development*, edited by James. C. Kaufman and
John Baer. Cambridge: Cambridge University Press.

Sallis, John. (2008). *Transfigurements: On the True Sense of Art*. Chicago, IL:
University of Chicago Press.

Sarbadhikari, Supten and Krishnendu Chakrabarty. (2001). 'Chaos in the Brain:
A Short Review Alluding to Epilepsy, Depression, Exercise and
Lateralization'. *Medical Engineering and Physics* 23 (7): 445–455. https://doi
.org/10.1016/S1350-4533(01)00075-3.

Sawyer, Keith. (2011). 'The Cognitive Neuroscience of Creativity: A Critical
Review'. *Creativity Research Journal* 23 (2): 137–154. https://doi.org/10.1080/1
0400419.2011.571191.

Schapiro, Meyer. (1937). 'Nature of Abstract Art'. *Marxist Quarterly* 1 (1):
77–98. www.on-curating.org/issue-20-reader/nature-of-abstract-art.html#
.XOy7y9Mza8o.

Schiffman, Harvey R. (2001). *Sensation and Perception: An Integrated Approach*.
New York: John Wiley & Sons.

Schimmel, Paul. (2012). 'Painting the Void'. In *Destroy the Picture: Painting the
Void, 1949–1962*, edited by Paul Schimmel, Nicholas Cullinan,
Astrid Handa-Gagnard, Shōichi Hirai, Sarah-Neel Smith, and
Robert Storr, 188–203. Paris: Rizzoli.

Schooler, Jonathan W., Jonathan Smallwood, Kalina Christoff, Todd C. Handy,
Erik D. Reichle, and Michael A. Sayette. (2011). 'Meta-Awareness,

Perceptual Decoupling and the Wandering Mind'. *Trends in Cognitive Sciences* 15 (7): 319–326. https://doi.org/10.1016/j.tics.2011.05.006.

Schooler, Jonathan W., Michael D. Mrazek, Michael S. Franklin, Benjamin Baird, Benjamin W. Mooneyham, Claire Zedelius, and James M. Broadway. (2014). 'The Middle Way: Finding the Balance between Mindfulness and Mind-Wandering'. In *The Psychology of Learning and Motivation: Volume 60*, edited by Brian H. Ross, 1–33. Waltham, MA: Academic Press.

Schopenhauer, Arthur. [1818] (1909). *The World as Will and Representation.* Translated and edited by Richard B. Haldane and John Kemp. London: Kegan Paul, Trench, Trübner & Co. www.gutenberg.org/ebooks/38427.

Selz, Peter and Jean Dubuffet (1962). *The Work of Jean Dubuffet.* New York, NY: Museum of Modern Art.

Shamay-Tsoory, Simon G., Noga Adler, Judith Aharon-Peretz, Daniella Perry, and Naama Mayseless. (2011). 'The Origins of Originality: The Neural Bases of Creative Thinking and Originality'. *Neuropsychologia* 49 (2): 178–185. https://doi.org/10.1016/j.neuropsychologia.2010.11.020.

Shaviro, Steven. (2012). *Without Criteria: Kant, Whitehead, Deleuze, and Aesthetics.* Cambridge, MA: MIT Press.

(2013). 'Non-Phenomenological Thought'. In *Speculations V: Aesthetics in the 21st Century*, edited by Ridvan Askin, Paul J. Ennis, Andreas Hägler, and Philipp Schweighauser, 40–56. Brooklyn, NY: Punctum Books.

(2018). 'Non-Correlational Thought'. In *Realism Materialism Art*, edited by Suhail Malik and Christoph Cox, 193–198. New York, NY: Sternberg Press.

Shulman, Gordon L., Maurizio Corbetta, Julie A. Fiez, Randy L. Buckner, Francis M. Miezin, Marcus E. Raichle, and Steven E. Petersen. (1997). 'Searching for Activations That Generalize over Tasks'. *Human Brain Mapping* 5 (4): 317–322. https://doi.org/10.1002/(SICI)1097–0193(1997) 5:4%3C317::AID-HBM19%3E3.0.CO;2-A.

Siegel, Jeanne. (1970). 'Carl Andre: Artworker'. *Studio International* 180 (927): 175–179.

Simondon, Gilbert. (1964). *L'individu et sa genèse physico-biologique.* Paris: Presses Universitaires de France.

Simony, Erez, Christopher J. Honey, Janice Chen, Olga Lositsky, Yaara Yeshurun, Ami Wiesel, and Uri Hasson. (2016). 'Dynamic Reconfiguration of the Default Mode Network during Narrative Comprehension'. *Nature Communications* 7: 12141. https://doi.org/10.1038/ncomms12141.

Smallwood, Jonathan, Daniel J. Fishman, and Jonathan W. Schooler. (2007). 'Counting the Cost of an Absent Mind: Mind Wandering as an Underrecognized Influence on Educational Performance'. *Psychonomic Bulletin and Review* 14 (2): 230–236. https://doi.org/10.3758/BF03194057.

Smithson, Robert. (1979). *The Writings of Robert Smithson.* Edited by Nancy Holt. New York, NY: New York University Press.

(1996). 'A Sedimentation of the Mind: Earth Projects'. In *Robert Smithson: The Collected Writings*, edited by Jack Flam, 82–91. Berkeley, CA: University of California Press.

Solomon R. Guggenheim Museum. (1981). *Jean Dubuffet: A Retrospective Glance at Eighty. From the Collections of Morton and Linda Janklow and the Guggenheim Museum, New York*. New York, NY: The Solomon R. Guggenheim Foundation. https://archive.org/details/jeandubuffetretr00newy.

Sreenivas, Shubha, Stephan G. Boehm, and David E. J. Linden. (2012). 'Emotional Faces and the Default Mode Network'. *Neuroscience Letters* 506 (2): 229–234. https://doi.org/10.1016/j.neulet.2011.11.012.

Starr, G. Gabrielle. (2013). *Feeling Beauty: The Neuroscience of Aesthetic Experience*. Cambridge, MA: MIT Press.

Stiles, Peter and Kristine Selz. (1996). *Theories and Documents of Contemporary Art*. Berkeley, CA: University of California Press.

Stoekl, Allan. (1985). 'Introduction'. In *Visions of Excess: Selected Writings, 1927–1939*, by Georges Bataille, ix–xxv. Minneapolis, MN: University of Minnesota Press.

Storr, Richard. (2012). 'Burnt Holes, Bloody Holes, Black Holes: Art after Catastrophe'. In *Destroy the Picture: Painting the Void, 1949–1962*, edited by Paul Schimmel, Nicholas Cullinan, Astrid Handa-Gagnard, Shōichi Hirai, Sarah-Neel Smith and Robert Storr, 240–257. Paris: Rizzoli.

Suzuki, Daisetz T. (1994). *An Introduction to Zen*. Foreword by C. G. Jung. New York, NY: Grove Press.

Sylvester, David. (1997). *About Modern Art: Critical Essays, 1948–1997*. New York, NY: Henry Holt.

Takahashi, Shigeko. (1995). 'Aesthetic Properties of Pictorial Perception'. *Psychological Review* 102 (4): 671–683. http://dx.doi.org/10.1037/0033-295 X.102.4.671.

Talmy, Leonard. (1996). 'Fictive Motion in Language and "Ception"'. In *Language and Space*, edited by Paul Bloom, Mary A. Peterson, Lynn Nadel, and Merrill F. Garrett, 211–276. Cambridge, MA: MIT Press.

Tàpies, Antoni. [1970] (2013). 'A Report on the Wall'. In *Tàpies From Within: 1945–2011*. Exhibition catalogue. Barcelona: Museu Nacional d'Art de Catalunya. Originally published in Antoni Tàpies, 'Comunicació sobre el mur', *La pràctica de l'art*, Barcelona: Ariel, 1970. www.museunacional.cat/s ites/default/files/tapies_-_dossier_de_premsa_eng.pdf.

Tàpies, Antoni, Anna Agustí, Georges Raillard, Miquel Tàpies, and Richard Lewis Rees. (1988). *Tàpies: The Complete Works*. New York: Rizzoli.

Taylor, Richard P., Branka Spehar, Paul Van Donkelaar, and Caroline M. Hagerhall. (2011). 'Perceptual and Physiological Responses to Jackson Pollock's Fractals'. *Frontiers in Human Neuroscience* 5: 60. https://doi.org/10 .3389/fnhum.2011.00060.

Thaut, Michael. (2005). *Rhythm, Music, and the Brain: Scientific Foundations and Clinical Applications*. London and New York, NY: Routledge.

The Times. (2008). 'La Loule du Virtuel by Jean Dubuffet'. 11 November 2008. www.thetimes.co.uk/article/la-loule-du-virtuel-by-jean-dubuffet-7jlv96g3nkc.

Thomson, David R., Brandon C. W. Ralph, Derek Besner, and Daniel Smilek. (2015). 'The More Your Mind Wanders, the Smaller Your Attentional Blink: An Individual Differences Study'. *Quarterly Journal of Experimental Psychology* 68 (1): 181–191. https://doi.org/10.1080/17470218.2014.940985.

Thomson, Iain D. (2011). *Heidegger, Art, and Postmodernity.* Cambridge: Cambridge University Press.

Tilley, Christopher. (2007). 'Materiality in Materials'. *Archaeological Dialogues* 14 (1): 16–20. https://doi.org/10.1017/S1380203807002139.

Tognoli, Emanuelle and J. A. Scott Kelso. (2014). 'The Metastable Brain'. *Neuron* 81 (1): 35–48. https://doi.org/10.1016/j.neuron.2013.12.022.

Toland, Alexandra and Gerd Wessolek. (2014). 'Picturing Soil: Aesthetic Approaches to Raising Soil Awareness in Contemporary Art'. In *The Soil Underfoot: Infinite Possibilities For a Finite Resource*, edited by G. Jock Churchman and Edward R. Landa, 83–102. Boca Raton, FL: CRC Press.

Tomasino, Barbara, Alberto Chiesa, and Franco Fabbro. (2014). 'Disentangling the Neural Mechanisms Involved in Hinduism- and Buddhism-Related Meditations'. *Brain and Cognition* 90: 32–40. https://doi.org/10.1016/j.bandc.2014.03.013.

Tomkins, Calvin. (1968). *The Bride and the Bachelors: Five Masters of the Avant Garde.* New York, NY: Viking Press.

Torre, Carlos A. (1995). 'Chaos, Creativity, and Innovation: Toward a Dynamical Model of Problem Solving'. In *Chaos Theory in Psychology and the Life Sciences*, edited by Robin Robertson and Allan Combs, 179–198. Marwah, NJ: Lawrence Erlbaum.

Travis, Fred and Jonathan Shear. (2010). 'Focused Attention, Open Monitoring and Automatic Self-Transcending: Categories to Organize Meditations from Vedic, Buddhist and Chinese Traditions'. *Consciousness and Cognition* 19 (4): 1110–1118. https://doi.org/10.1016/j.concog.2010.01.007.

Tsuda, Ichiro. (2001). 'Toward an Interpretation of Dynamic Neural Activity in Terms of Chaotic Dynamical Systems'. *Behavioral and Brain Sciences* 24 (5): 793–810. https://doi.org/10.1017/S0140525X01000097.

Tsuda, Ichiro, Yoko Yamaguchi, Takashi Hashimoto, Jiro Okuda, Masahiro Kawasaki, and Yasuo Nagasaka. (2015). 'Study of the Neural Dynamics for Understanding Communication in Terms of Complex Hetero Systems'. *Neuroscience Research* 90 (January): 51–55. https://doi.org/10.1016/j.neures.2014.10.007.

Tucker, Marcia. (1973). *Lee Krasner: Large Paintings.* New York, NY: Whitney Museum of American Art.

Van Leeuwen, Cees. (1998). 'Visual Perception on the Edge of Chaos'. In *System Theories and A Priori Aspects of Perception*, edited by J. Scott Jordan, 289–314. Amsterdam: Elsevier.

(2007). 'What Needs to Emerge to Make You Conscious?' *Journal of Consciousness Studies* 14, no. 1–2: 115–136. www.ingentaconnect.com/content/imp/jcs/2007/00000014/f0020001/art00008.

Van Leeuwen, Cees and Dirk J. A. Smit. (2012). 'Restless minds, Wandering Brains'. In *Being in Time: Dynamical Models of Phenomenal Experience*, edited by Shimon Edelman, Tomer Fekete, and Neta Zach, 121–148. New York, NY: John Benjamins.

Varnedoe, Kirk. (2006). *Pictures of Nothing: Abstract Art Since Pollock.* Princeton, NJ: Princeton University Press.

Varnedoe, Kirk and Pepe Karmel. (1999). *Jackson Pollock: New Approaches*. New York, NY: Museum of Modern Art.

Vartanian, Oshin. (2018). 'Internal Orientation in Aesthetic Experience'. In *The Oxford Handbook of Spontaneous Thought: Mind-Wandering, Creativity, and Dreaming*, edited by Kieran C. R. Fox and Kalina Christoff, 327–336. Oxford: Oxford University Press.

Vessel, Edward A., G. Gabrielle Starr, and Nava Rubin. (2012). 'The Brain on Art: Intense Aesthetic Experience Activates the Default Mode Network'. *Frontiers in Human Neuroscience* 6: 66. https://doi.org/10.3389/fnhum.2012.00066.

(2013). 'Art Reaches Within: Aesthetic Experience, the Self and the Default Mode Network'. *Frontiers in Neuroscience* 7: 258. https://doi.org/10.3389/fnins.2013.00258.

Wagner, Monica. (2015). 'Material'. In *Materiality*, edited by Petra Lange-Berndt, 26–29. London: Whitechapel Gallery, and Cambridge, MA: MIT Press.

Wallas, Graham. (1945). *The Art of Thought*. London: C.A. Watts & Co., Ltd.

Wang, Tingting, Lei Mo, Oshin Vartanian, Jonathan S. Cant, and Gerald Cupchik. (2015). 'An Investigation of the Neural Substrates of Mind Wandering Induced by Viewing Traditional Chinese Landscape Paintings'. *Frontiers in Human Neuroscience* 8: 1018. https://doi.org/10.3389/fnhum.2014.01018.

Watanabé, Satosi. (1969). 'The Concept of Time in Modern Physics and Bergson's Pure Duration.' In *Bergson and the Evolution of Physics*, edited and translated by Pete Addison Gunter, 62–76. Knoxville, TN: University of Tennessee Press.

Weir, Catherine and Evans Mandes. (2017). *Interpreting Visual Art: A Survey of Cognitive Research About Pictures*. Piscataway, NJ: Transaction Publishers.

Westgeest, Helen. (1997). *Zen in the Fifties: Interaction in Art Between East and West*. The Hague: Waanders Publishing.

Wicks, Robert. (2013). *Kant: A Complete Introduction*. Boston, MA: Hachette.

Will, Udo and Gabe Turow. (2011). 'Introduction to Entrainment and Ethnomusicology'. In *Music, Science, and the Rhythmic Brain: Cultural and Clinical Implications*, edited by Jonathan Berger and Gabe Turow, 3–30. New York, NY: Routledge.

Williamson, Beth. (2017). *Between Art Practice and Psychoanalysis Mid-Twentieth Century: Anton Ehrenzweig in Context*. London: Routledge.

Wilson, Sarah. (2013). 'Bergson before Deleuze. How to Read *Informel* Painting'. In *Bergson and the Art of Immanence: Painting, Photography, Film*, edited by John Mullarkey and Charlotte de Mille, 80–93. Edinburgh: Edinburgh University Press.

Wilson, Stephen. (2010). *Art + Science Now*. London: Thames & Hudson.

Wilson-Mendenhall, Christine D., Lisa Feldman Barrett, W. Kyle Simmons, and Laurence W. Barsalou. (2011). 'Grounding Emotion in Situated Conceptualization'. *Neuropsychologia* 49 (5): 1105–1127. https://doi.org/10.1016/j.neuropsychologia.2010.12.032.

Winawer, Jonathan, Alexander C. Huk, and Lera Boroditsky. (2010) 'A Motion Aftereffect from Visual Imagery of Motion'. *Cognition* 114 (2): 276–284. https://doi.org/10.1016/j.cognition.2009.09.010.

Wiskus, Jessica. (2013). *The Rhythm of Thought: Art, Literature, and Music After Merleau-Ponty*. Chicago, IL: University of Chicago Press.

Worringer, Wilhelm. [1908] (1997). *Abstraction and Empathy: A Contribution to the Psychology of Style*. Translated by Michael Bullock. Chicago, IL: Elephant/Ivan R. Dee.

Yang, Jiongjiong, Xuchu Weng, Yufeng Zang, Mingwei Xu, and Xiaohong Xu. (2010). 'Sustained Activity within the Default Mode Network during an Implicit Memory Task'. *Cortex* 46 (3): 354–366. https://doi.org/10.1016/j.cortex.2009.05.002.

Yarbus, Alfred L. (1967). *Eye Movements and Vision*. New York, NY: Plenum Press.

Yoshihara, Jirō. [1956] (1996). 'The Gutai Manifesto'. In *Theories and Documents of Contemporary Art: A Sourcebook of Artists' Writings*, edited by Kristine Stiles and Peter Selz, 695–698. Berkeley, CA: University of California Press. http://web.guggenheim.org/exhibitions/gutai/data/manifesto.html.

Zajonc, Robert B. (1980). 'Feeling and Thinking: Preferences Need No Inferences'. *American Psychologist* 35 (2): 151–175. http://dx.doi.org/10.1037/0003-066X.35.2.151.

Zangemeister, Wolfgang H., Keith R. Sherman, and Lawrence W. Stark. (1995). 'Evidence for a Global Scanpath Strategy in Viewing Abstract Compared with Realistic Images'. *Neuropsychologia* 33 (8): 1009–1025. https://doi.org/10.1016/0028-3932(95)00014-T.

Zeki, Semir. (1999). *Inner Vision: An Exploration of Art and the Brain*. Oxford: Oxford University Press.

Index